The political puzzle of climate change is not only about why so little has been accomplished internationally, it is also how climate politics has been subsumed into the discourse and practices of climate security. In an acute analysis, Delf Rothe examines both the nuts and bolts of this matter and the underlying structural processes that have turned an already complex issue into an even more wicked problem.

Ronnie D. Lipschutz, *University of California, Santa Cruz, USA*

This book is an ambitious attempt to examine current theories of securitization and to rethink the links between climate and security in contemporary policy discourse. Focusing on the United Kingdom in particular this analysis of the rise of climate risks and biopolitical security policy moves academic debates ahead usefully. Essential reading for anyone interested in understanding contemporary developments in environmental security thinking and policy innovation.

Simon Dalby, *Balsillie School of International Affairs, Wilfrid Laurier University, Canada*

Securitizing Global Warming

This book explores the reasons for the recent securitization of climate change, and reveals how the understanding of climate change as a security threat fuels resilience as a contemporary political paradigm.

Since 2007, political and public discourse has portrayed climate change in terms of international or national security. This increasing attention to the security implications of climate change is puzzling, however, given the fact that linkages between climate change and conflict or violence are heavily disputed in the empirical literature. This book explains this trend of a securitization of global warming and discusses its political implications. It traces the actor coalition that promoted the idea of climate change as a security issue and reveals the symbols, narratives and storylines that make up this discourse. Drawing on three detailed case studies at the international level of the United Nations, the regional level of the Euro-Mediterranean and the national level of the UK, the book reveals how climate change is turned into a non-linear and unpredictable threat. The resulting complexity discourse prevents the adoption of any exceptional measures and instead presents resilience as the only way to cope with the climate threat. This book shows that we can only grasp the complexity of the securitization process and its implications in the climate change case by comparing it at different political levels over a longer period. By developing a securitization framework, the book makes an important contribution to the ongoing debate on security and resilience in critical security studies.

This book will be of much interest to students of critical security studies, resilience, environmental studies, global governance and IR in general.

Delf Rothe is a Postdoc-Fellow at the Institute for Peace Research and Security Policy in Hamburg, Germany, and has a PhD in political science.

Routledge Studies in Resilience
Series Editor: David Chandler
University of Westminster

The *Routledge Studies in Resilience* series is interested in publishing a broad range of high-quality contemporary research into the processes, spaces, policies, practices and subjectivities through which resilience is seen to operate.

Securitizing Global Warming
A climate of complexity
Delf Rothe

Securitizing Global Warming

A climate of complexity

Delf Rothe

Routledge
Taylor & Francis Group
LONDON AND NEW YORK

First published 2016
by Routledge
2 Park Square, Milton Park, Abingdon, Oxon OX14 4RN

and by Routledge
711 Third Avenue, New York, NY 10017

First issued in paperback 2017

Routledge is an imprint of the Taylor & Francis Group, an informa business

British Library Cataloguing-in-Publication Data
A catalogue record for this book is available from the British Library

Library of Congress Cataloging-in-Publication Data
Names: Rothe, Delf, author.
Title: Securitizing global warming : a climate of complexity / Delf Rothe.
Description: Abingdon, Oxon : New York, NY : Routledge is an imprint of the Taylor & Francis Group, an Informa business, [2016] | Series: Routledge studies in resilience | Includes bibliographical references and index.
Identifiers: LCCN 2015024485| ISBN 9781138935303 (hardback) | ISBN 9781315677514 (ebook)
Subjects: LCSH: Global warming–Political aspects. | Climatic changes–Political aspects. | National security–Environmental aspects. | Security, International–Environmental aspects.
Classification: LCC QC903 .R677 2016 | DDC 363.738/74–dc23
LC record available at http://lccn.loc.gov/2015024485

ISBN 13: 978-1-138-09653-0 (pbk)
ISBN 13: 978-1-138-93530-3 (hbk)

Typeset in Baskerville
by Wearset Ltd, Boldon, Tyne and Wear

For Lotta

Contents

Figures

Tables

Acknowledgments

In the course of writing this book I have worked together with a lot of people that helped in shaping the project. Not to mention and thank them would hence be an outrageous injustice. First of all, this book would not have come together without the support of the *Heinrich-Böll-Foundation*, which funded my PhD project at the University of Hamburg from April 2009 until March 2012. Far beyond material support, they've done everything to provide the backing that is necessary to successfully conduct a dissertation project. Moreover, I am much obliged to my PhD supervisor *Professor Cord Jakobeit* for his trust in my project and his always-helpful advice. I want to thank *Chris Methmann* and *Benjamin Stephan* for being such good friends and colleagues. If I didn't mention it before: your comments and critique have been invaluable for writing this book. Many thanks also to *Professor Annette Jünemann*: it has always been a pleasure to work for and with you over the past – already nine – years. *Professor Marieke de Goede* as well as *Marijn Hoijtink*, *Stephanie Simon* and *Gavin Sullivan* did everything to make my stay at the University of Amsterdam from January until March 2012 pleasant, productive and successful. Thank you for inspiring discussions, helpful comments and nice evenings in Amsterdam. *Florian Kühn*, *Angela Oels*, *Franz von Lucke Giovanni Bettini* as well as *Jakob Horst* all made time for reading and commenting on chapters of this book. Your ideas, arguments and suggestions have been very helpful and I think they raised the quality of this study considerably. *Fiona Woo* took on the hard job of proofreading my English writing. Many thanks for not going crazy and for your very thorough work.

I owe special thanks to my partner *Neele Montag* for always listening to my complaints and worries, for calming me down in moments of crisis and for motivating and encouraging me. Without your help I would not have been able to realize my project as I did with your support. Last but not least, I want to thank my parents, *Jutta Rothe* and *Wolfgang Rentzel-Rothe*, who always believed in me and supported me whenever and however they could.

Abbreviations

CCIS	Climate change and international security
CDM	Clean Development Mechanism
CDP	Carbon Disclosure Project
CIM	Climate-induced migration
CIRAM	Climate Impact Risk Assessment Method
CSDC	Climate security discourse coalition
DCDC	Development, Concepts and Doctrine Centre (UK)
DECC	Department of Energy and Climate Change
Defra	Department for Environment, Food and Rural Affairs (UK)
DfID	Department for International Development
DSLD	Desertification, and soil and land degradation
EMP	Euro-Mediterranean Partnership
ENP	European Neighbourhood Policy
ESS	European Security Strategy
FCO	Foreign and Commonwealth Office
G77	The Group of 77
GECHS	Global Environmental Change and Security Project
ICF	International Climate Fund
IOM	International Organisation for Migration
IPCC	Intergovernmental Panel on Climate Change
MedSec	Mediterranean Security Partnership
MENA	Middle East and North Africa
MoD	Ministry of Defence
NATO	North Atlantic Treaty Organisation
NSS	National Security Strategy
ODA	Official Development Assistance
OSCE	Organisation for Security and Cooperation in Europe
PPRD	Prevention, Preparedness and Response to Natural and Man-Made Disasters
SES	Socio-Ecological System
SIDS	Small Island Development States
UfM	Union for the Mediterranean
UN	United Nations

UNDP	United Nations Development Programme
UNEP	United Nations Environment Programme
UNFCCC	United Nations Framework Convention on Climate Change
UNSC	United Nations Security Council
UNGA	United Nations General Assembly
WBGU	German Advisory Council on Global Change

1 Introduction – or how discourses create a climate of complexity

> We, the human species, are confronting a planetary emergency – a threat to the survival of our civilization that is gathering ominous and destructive potential even as we gather here. [...] without realizing it, we have begun to wage war on the earth itself. Now, we and the earth's climate are locked in a relationship familiar to war planners: "Mutually assured destruction." [...] We must quickly mobilize our civilization with the urgency and resolve that has previously been seen only when nations mobilized for war.
>
> (Al Gore in his Nobel Peace Prize lecture 2007)

In 2007, the issue of global climate change received an unprecedented amount of public and political attention, which culminated in the awarding of the Nobel Peace Prize to Al Gore and the Intergovernmental Panel on Climate Change (IPCC). The more scientific findings of the IPCC were spread by mass media worldwide, the more climate change advanced to become an issue of high politics. At the same time, political and public discourses on climate change took up a new framing or storyline that portrayed global warming in terms of (inter-)national security. Over the following years arguments linking the issue of climate change to either national or international security politics were brought forward by prominent politicians, such as the then UK Foreign Minister Margaret Beckett in 2007 (Beckett 2007), the European Union's High Representative for Common Foreign and Security Policy Javier Solana in 2008 (Solana 2008), or the United Nations' (UN) Secretary General Ban Ki-moon (e.g. Ki-Moon 2009) in 2009. Important political bodies in global security governance – such as the United Nations Security Council (UNSC) in 2007 and 2011 as well as the United Nations General Assembly (UNGA) in 2009 – have discussed the security implications of climate change and possible political means to deal with them (see UNGA 2009; UNSC 2007a, b, 2011a, b).

The increasing public and political attention to the security implications of climate change is puzzling, however, given the fact that robust scientific knowledge on the empirical correlation of climatic changes and the onset of violent conflicts was scarce at that time – and remains so at present. Summarizing the vast academic debate, one could state that while some see climate change as the source of 'climate wars' and novel kinds of conflicts (Dyer 2009; Fetzek and Mazo 2014) or even as a hyper-variable for global security in the twenty-first century (see Smith 2007: 832), others severely doubt the explanatory power of the variable 'climate change' for the occurrence of violent conflicts and other security threats (e.g. Brown *et al.* 2007; Buhaug *et al.* 2014; Theisen *et al.* 2012).

Some political commentators and academic scholars went so far as to not only criticize the empirical basis of climate security discourses but to explicitly warn against their potential negative implications (e.g. Buckland 2007). These critical voices warned that discussing climate change in terms of security and survival might prompt the wrong policy measures. Some have argued that a security take on climate change bears the risk of an increasing militarization of climate politics and could strengthen military/ security actors compared to environmental agencies (Detraz and Betsill 2009: 314). The employment of the US military in the aftermath of Hurricane Katrina (Smith 2007: 332–333) or the deployment of US troops in African crisis regions as part of the US military's Africa Command 'USAFRICOM' (Hartmann 2014: 775) have been cited as examples of this tendency. Securitizing climate change, i.e. framing it as an existential security threat, it was feared, could moreover influence the decision-making process in international climate governance. For example, it could shift political authority from inclusive- and consensus-oriented institutions such as the United Nations Framework Convention on Climate Change (UNFCCC) to more exclusive political bodies (Reiber and Zelli 2011: 220). The UNSC, which has debated the issue of climate change twice since 2007, would be a good example of such an exclusive political body. So is the G7, which announced in April 2015 that it considered the issue of "Climate change, security and fragility as a foreign policy priority" (Climateandsecurity.org 2015) Lastly, it has been argued that climate security arguments might even justify the adoption of extreme measures such as geoengineering in climate governance (Dalby 2015: 14; Fleming 2007: 21).

The rise of climate security discourse and the resilience conundrum

In this book I thus set out to investigate a twofold puzzle. On the one hand, since 2007 a security framing of climate change became ever more prominent in political discourse although scientific facts about the security implications of anthropogenic climate change are far from reliable (Dalby 2015: 13; Hartmann 2014: 779). On the other hand, we did not see the just mentioned negative policy implications that have been associated with such a discursive shift (Brzoska 2012: 58–59; von Lucke *et al.* 2014: 2). Even though the notion of global warming as a major security threat got accepted by an increasing number of political actors, global climate governance continued to be a piecemeal, technocratic and often messy endeavor (e.g. Detraz and Betsill 2009: 311). The release of a 'Climate Change Adaptation Roadmap' by the US Pentagon in October 2014 is a good case in point. Presenting the report to the public on October 13, 2014 US Secretary of Defense Chuck Hagel stressed that climate change poses a threat both to international and to US national security (Lamothe 2014). However, it was not as if this securitizing move was followed by a call for a deployment of US troops in climate conflict areas, or for a more progressive US policy to mitigate the risk of climate change. Also the report does not make any concrete budget recommendations for implementing the suggested policies (Davenport 2014). In short, there is no sign that the military or the defense ministry seeks to position itself as an actor in climate politics or to seize the issue from the environmental department. Rather, the 'Climate Change Adaptation Roadmap' (Pentagon 2014) outlines a technical *approach of resilience* to prepare the US army for the impacts of climate change through risk assessments, training exercises, or infrastructure investment.

In short a resilience approach to climate security (which will be further described in Chapters 5–7) focuses on the self-capacities of communities and socio-natural systems to cope with environmentally induced shocks and to adapt to changing environmental conditions. Resilience stresses learning-based, reflexive governance mechanisms in order to be able to deal with uncertainties and contingencies. Instead of top-down government interventions, it preaches flexibility, empowerment and inclusion. Whereas one might expect that a security take on climate change would reinforce the role of (inter-)national actors in climate politics, the opposite seems to be the case: the governance of climate risks is being dispersed among a whole array of public and private actors including communities, cities, NGOs and businesses. And while it had been feared a securitization would lead states to adopt extreme measures, such as shutting down borders to ward off climate refugees or turning to geoengineering to modify climate patterns, the actual policies to deal with climate insecurity are completely mundane. Such activities include funding local adaptation projects, raising awareness and fostering learning (see Chapter 5). Climate security policies seem to be dominated by a logic of resilience that fundamentally opposes the alarmist tone informing climate security discourse.

This book is thus guided by two broad research questions each addressing one side of the outlined puzzle. First, I am asking: *How and why has climate change been discursively turned into a security problem despite the scientific uncertainties about the climate–security link?* Second, I address the resilience conundrum and ask for the reasons for the apparent misfit between the climate security discourse and the policies that it promotes. *How can it be that climate security discourse with its alarmist, exceptional, sometimes even apocalyptic rhetoric leads into the mundane practices of promoting resilience?* In this book I try to answer these deliberately open questions, allowing complexity and surprise in the securitization of climate change without narrowing it down to a particular security discourse.

Studying the securitization of climate change

There is one theoretical approach that immediately comes to mind when it comes to the extension of the security sector to new issue areas, such as that of climate change: this is the securitization approach of the so called Copenhagen School. In search of a theoretical framework to address the twofold puzzle just outlined, the Copenhagen School thus should be the first choice. As I show in this section, while the securitization framework of the Copenhagen School is perfectly equipped to explain the first part of the puzzle, it fails to explain the second: that is the mundane logic of resilience that undergirds practices of climate insecurity. The main assumptions of the Copenhagen School as well as the reasons for its failure in the climate change case are summarized in the following.

The general assumption of the Copenhagen School is that security problems do not just exist out there but are socially constructed. Buzan *et al.* substantiate this claim by drawing on linguistic speech act theory (Austin 1975; Searle 1969), which emphasizes the *performative* – that is, *pragmatic character* – of language. Language for them is more than a medium to exchange information – it represents a form of social action. Following speech act theory, the articulation of a security problem should not be seen as a mere description of reality. Rather, if political actors speak about security, they actively construct security problems. Security, then, can be understood as a social construction of a particular reality based on the performative force of language. Securitization takes place through speech acts that follow a particular conventionalized – that is, socially agreed upon – script: it identifies a particular phenomenon as an existential threat to a valued reference object (Buzan *et al.* 1998: 26). This speech act or securitizing move takes an issue beyond the established rules of the game. The urgency and exceptionality that characterize security problems legitimize the adoption of political measures that would not be possible in 'normal' politics (Wæver 1995: 55). By following the approach of the Copenhagen School we would hence have to assume that a securitization process takes the issue of climate change out of the established rules of normal politics and subjects it to the exceptional mode in which established political rules of the given polity do not apply. However, as I have outlined above, in the case of climate governance we are completely lacking any exceptional or extreme measures.

A first and most obvious explanation for this twofold puzzle could be that we are dealing with a *failed securitization* in the climate change case. The Copenhagen School's approach assumes that a securitizing move can fail if it does not manage to convince a relevant audience (Buzan *et al.* 1998: 30–32). In the case of climate change, however, the possibility of a failed securitization can be ruled out for two reasons. First, there is broad consensus on the security implications of climate change, which is shared by a wide range of different political actors – from national leaders to think tanks and environmental or humanitarian NGOs. While articulations that linked climate change to the occurrence of violent conflict have often triggered public attention,[1] critical voices were seldom disseminated, as my analysis of international media coverage proves. The fact that the UK Ministry of Defence (MoD 2010) and the US Pentagon (2014) have mainstreamed climate adaptation into their operations clearly shows that climate security discourse found an echo from important actors in the security field. Second, if one takes a close look at the proposed policy measures within climate security discourse, one can recognize that none of them lobbies for the adoption of any exceptional measures. Rather, all promoted political measures refer to everyday and very mundane practices, for example small-scale technical adaptation projects to increase the resilience of vulnerable communities (Methmann and Rothe 2013).

If the possibility of a failed securitization is ruled out, it follows that there must be something special about the climate change case, which poses a challenge to the Copenhagen School's framework of securitization. First and foremost, climate change displays a distinct temporal order that differs from that of other recently securitized issues such as international migration, radical Islamism or Ebola. Climate change poses *future risks* – meaning that its negative security implications are not yet felt today but will only (potentially) be experienced within an uncertain time span of the next 20–100 years – the whole discourse is based on different forms of future-conditional knowledge (Baldwin 2014: 213).

Furthermore, different studies have shown that the securitization of climate change is not reducible to single securitizing acts but takes the form of an ongoing process, influenced by the practices of very different actors ranging from climate scientists, through environmental activists and political celebrities to security professionals and practitioners (Brzoska 2012). In other words, the concrete relation between climate change and security in these discourses is not a neutral, given one but is itself subject to contestation. This observation poses a challenge to the model of the Copenhagen School which concentrates on single securitizing moves of central, authoritative actors (McDonald 2008). At the same time, it was shown that the discourse was characterized by many different concepts of security (Brzoska 2009) and hence a concentration on a realist or military centered notion of security would be counterproductive (von Lucke *et al.* 2014: 3). This coincides with the findings of scholars such as Julia Trombetta or Michael Brzoska and Angela Oels, which state that in the case of climate change one should turn the perspective of the Copenhagen School upside down and speak of a climatization of security and development policy (Oels and

Brzoska 2012; Trombetta 2008). Following them we can notice that concepts from an ecological context such as sustainability and resilience are increasingly adopted and reinterpreted by actors in the fields of development and security policy (see Chapter 7). Along the same lines, different scholars have argued that the securitization of climate change took place mainly in the register and language of *risk* and resilience – and hence points to a blind-spot of the Copenhagen School, which is the ignorance of risk discourses and practices, i.e. security discourses *below the threshold of existentialism* (Corry 2014; Methmann and Oels 2015).

In this book I set out to solve the twofold puzzle outlined above by critically re-examining the securitization account of the Copenhagen School against the backdrop of the climate change case. The book provides an in-depth and detailed study of security discourses and their political implications at three different political levels – the international, the regional and the national – and over a longer period. I account for both the recent peak phase of climate security discourse from 2007 until 2012 as well as the longer genealogy of this discourse since the early 1990s. By doing so I do not only account for the tip of the iceberg – i.e. the most visible and prominent articulations of climate security linkages – but also for the lower lying, partly concealed discursive foundation of such statements. Lastly, by drawing on a broad notion of discourse I extend the analytical perspective beyond speech acts (i.e. linguistic articulations) to include different forms of practices (non-linguistic articulations of climate security) as well as their institutional backgrounds.

Selection of the cases

To paint a more comprehensive picture of the securitization of climate change and its political implications, this study broadens the perspective to include different political levels and different policy fields. To this end, I distinguish three cases, each of which focuses on a different political scale and policy field of the climate security discourse: the discourse at the international level of the UN, the European discourse on the Mediterranean as a climate security hotspot, and the national discourse within the United Kingdom (UK). To be precise: there is only one climate change and security discourse, which does not stop or change at national or regional borders.[2] However, depending on the respective context, the discourse takes different forms and has different political implications. In this study each context is investigated through an in-depth analysis, allowing one to compare different forms and manifestations of the securitization process that I conceptualize as different logics. As Jackson notes, interpretive or reflexive researchers "have no profound epistemic need for comparison, although when they do compare cases they do so for quite distinct purposes: […] to 'incorporate' multiple cases into a more holistic view that supervenes on the individual case" (Jackson 2010: 200). Hence, comparison here is not understood as a method to infer a particular type of knowledge (Stake 1994: 237) – e.g. generalizable causal mechanisms. Rather, comparison on the one hand aims to reveal contradictions and fissures and on the other hand searches for similarities between the discursive levels. Together this allows for the painting of a more comprehensive picture of the securitization process.

To produce such a comprehensive picture I drew on what I have called a *retroductive comparison*. The basic idea is to fuse in-depth study with practices of comparison and theory development into one retroductive, i.e. circular, research process. The concept of *retroduction* refers to a research posture which is neither descending nor ascending but, rather, circular or oscillating between theses formulation and testing. The starting point in a *retroductive circle* is the identification of a *puzzle* – an anomalous or wondrous phenomenon. This will incite the investigator to develop some first hypotheses or proto-explanations that could explain the puzzle (Glynos and Howarth 2007: 35). The subsequent 'testing' of these proto-explanations is an open and flexible process, which includes empirical analysis, just like critical and theoretical reflections, and practices of argumentation and persuasion (Glynos and Howarth 2007: 38). The research process does not end with an empirical investigation but includes processes of theoretical re-adjustment, revisions and adaptation of the early proto-explanations to novel findings and a continued extension of the empirical material.

The merit in having multiple circles of empirical 'testing' is that you can transfer concepts that were identified in one case to the other two levels in the next step of the retroductive circle. The findings in one case might incite you to reconsider your theoretical framework or to add further theoretical concepts, which then in a next step provide substantial insight in another case – and which were missed in an earlier stage of the retroductive circle.[3] To give an example, an explorative analysis of climate security discourse at the UN level revealed the important role of risk and resilience thinking in this context. This incited consideration of risk as a crucial theoretical element of a securitization approach. And this revised theoretical approach was then re-applied to the other cases as well in a next step of analysis to check whether this theoretical adaptation proved relevant in these contexts as well.

It is important to mention this circular procedure at this point, prior to the presentation of empirical results, as the linear structure of a book prohibits a circular presentation of results. My central aim of the retroductive comparison has been to carve out different logics of securitization. Some of these logics showed up across the different discursive levels, while others are peculiarities of a particular case.

All three different cases are to be considered as ideal types, which in empirical reality may imply overlap. Furthermore, as a way to avoid overlap and analytical inaccuracies, the analysis assigns articulations to one particular case on the basis of content and not on the basis of where the speakers come from.[4] All of the selected cases are characterized by recent discursive shifts, i.e. the reinforcement of security narratives in political debate (see Brauch 2012: 719–720; Oels and Brzoska 2012: 188–189). They can thus be considered as most likely crucial cases (Flyvbjerg 2001: 78–79) with regard to a securitization of climate change. Taken together, they are suitable for generating detailed insights into the link between security discourses on climate change and political action.

All of the three case studies seek to give answer to the core research questions of this book – the questions of how climate change turned into a security issue and the political implications. However, in order to reveal different logics of securitization in each case, I ask these questions in a slightly different way. In the first case study, I trace the climate change and security discourse in the context of the United Nations. My analytical focus here is on the process of discursive struggle and the emergence of a discourse coalition around the notion of climate security. In terms of the policy implications I take a closer look at the international climate regime UNFCCC and related international policy initiatives. In the second case, I turn my focus to a second policy field, i.e. EU development policy and practices of region building in the Mediterranean. I am particularly interested in the mechanisms of a discursive dissemination, i.e. the ways in which European actors have tried to export climate security rationales and related practices to their southern Mediterranean neighbors. Finally, I use the case of the UK to study how rationales and concepts of climate security discourse have trickled down into governmental ministries and organizations. Asking for the political implications of this process, I take a closer look at foreign and security policy – starting from the assumption that we might witness a 'climatization' of the security field, i.e. that the most drastic policy changes are to be observed in the security field itself rather than in environmental policy.

Table 1.1 Selection and differentiation of the cases

	Context 1	*Context 2*	*Context 3*
Political scale	International (the United Nations)	Regional (the discourse on the Mediterranean)	National (United Kingdom)
Policy field	Global climate governance	Development policy; region building	Foreign and security policy
Analytical focus on	Processes of discursive struggle	Practices of discursive dissemination	Institutionalization and policy implications of the discourse

Outline of the book

To find an answer to these questions, the second and third chapters of this book provide a thorough revision of the securitization framework in order to develop a *consequent poststructuralist theory of securitization.* The conceptual debate on securitization in the literature has been characterized by an apparent schism between the (language-)philosophical approach of the Copenhagen School and a sociological variant usually labeled as the Paris School (Chapter 2). The aim of this book is decidedly not to provide another variant of securitization theory or to engage in the debate between these two apparently opposing camps. Rather, the aim is to revise the framework of the Copenhagen School from within in order to strengthen its poststructuralist elements. By doing so, I argue, many of the apparent contradictions between a philosophical and a sociological version of securitization – between discourse and practice – dissolve.

To do so, I draw upon the method of deconstruction in order to reveal the conceptual building blocks and constitutive antagonisms within the securitization framework. The rationale behind this approach is that the Copenhagen School, by combining the thoughts of Carl Schmitt, Hans Morgenthau, John Austin's speech act theory as well as Jacques Derrida, among others, produces a series of contradictions and antinomies but also systematically suppresses the 'poststructuralist' thoughts it draws upon. In Chapter 3 I revisit the different elements of the framework in order to discuss possible contradictions and shortcomings. These elements are: the concept of discourse, the underlying notion of power, security as exception versus the politics of resilience, securitization processes and the political, and non-lingual processes of securitization. The aim of the chapter is to lay open the oppositional assumptions informing each conceptual building block, to discuss their epistemological and ontological roots, and to suggest means to their dissolution. To do so, I draw on the insights of a range of poststructuralist thinkers including most importantly Ernesto Laclau and Chantal Mouffe, Jacques Derrida and Michel Foucault. The result is a consequent poststructuralist theory of securitization. Based on these revisions I subsequently carve out my own poststructuralist approach of securitization, presented as a synthesis at the end of the chapter.

Without giving too much away at this point, let me briefly sketch the contours of such a consequent poststructuralist approach of securitization, which tries to tear down the defense walls between a philosophical and a sociological camp of securitization theory:

First, this revised securitization framework draws on the comprehensive understanding of discourse of Laclau and Mouffe (1985) in order to overcome the false dichotomy between discourse and practice or the ideational/material that haunts many debates on securitization. Instead of linguistic speech acts, my account is modeled along the lines of the notion of discourse coalitions (Hajer 1995).

Second, the growing importance of resilience and risk in contemporary security policy shows that the model of the Copenhagen School has been stretched to its limits. Hence, I open my approach to securitization to include different and opposing security concepts. However, in order to maintain conceptual clarity, I distinguish between certain logics of securitization. Drawing on a Foucauldian understanding of security, I finally show that a logic of defense, or geopolitical security, and a logic of risk/resilience, or biopolitical security, are two sides of the same coin.

Third, I argue that a binary distinction between politics and security is problematic. Drawing on Laclau and Mouffe's notion of the political, I instead develop a gradual notion of de- and re-politicization.

Chapter 4 tries to dig deeper into climate change and security discourse in order to trace its conceptual roots over time as well as to reveal its underlying conceptual deep structure. Presenting the results of a diachronic discourse analysis, I follow the discursive lineages of climate security back into the early 1990s and identify different moments of dislocation or turning points in the development of a genuine climate security discourse. The genealogical study also discusses how failure and contradictions as creative forces helped bring about climate security as a political matter of concern. In the second part of the chapter, I outline the discursive deep structure of climate security discourse as a result of a fine-tuned synchronic discourse analysis.

To paint a more comprehensive picture of the securitization of climate change and its political implications, Chapters 5 to 7 present the three case studies. Chapter 5 focuses on the climate change and security discourse at the *international level*. This case allows for the studying of how climate change and security could grow into an issue of international concern debated in major bodies of the UN such as the UNSC and UNGA. There is one fact that makes it a particularly strong case for studying the impact of a securitization process for *global climate governance*: much of the discourse on climate change and security has been staged at the UN and other bodies of global governance. The case thus allows the main actors, storylines and discursive strategies in climate change discourse to be studied.

Chapter 6 focuses on a regional level and studies the European Union's discourse on the security implications of climate change in its southern neighbourhood: the Mediterranean. The Mediterranean is usually regarded as one of the regions most threatened by a dangerous climate change (Brauch 2012: 726–749). Given its proximity to the European Union, one of the leading actors in the climate security discourse, it is a hot topic in the discourse on climate change and security as well. Hence, it represents a paradigmatic case with which to study the implications of such a discourse at a regional political scale. At the same time the case allows the political implications of the discourse in another policy field to be studied – namely, European development policy or region-building practices. While climate change and security discourse is widespread within the EU, this does not hold true for the southern Mediterranean region. The Euro-Mediterranean, hence, represents a crucial case to study the practices through which European actors try to disseminate and spread storylines of climate change and security.

Chapter 7 discusses the political and institutional effects of climate security discourse against the backdrop of the case of *the UK*. The United Kingdom represents one of the forerunners in the climate change and security debate (Boas 2015). Furthermore, a lot of the political implications can only manifest themselves at the national level – e.g. due to the state's monopoly of military force. In order to study these political impacts, I trace how actors from different policy fields – i.e. foreign, development and security policy – have adopted and implemented climate security rationales. I show how a logic of *climate security mainstreaming* prohibits the development of a genuine UK climate security policy. Instead, we can observe the emergence of a complex climate security assemblage, which is organized by the malleable and vague concept of resilience. In the final chapter, I conclude by comparing the different cases, summarizing the core arguments, and discussing the contribution of my research to the literature.

Notes

1 This holds true, for instance, for statements linking the Darfur conflict in Sudan to the effects of climate change. Another recent example was a study postulating a correlation between El Niño-related extreme weather and the probability of violent conflict. Both arguments received attention from major international newspapers (e.g. Bloomfield 2007; Fears 2011).

2 For example, there are actors in the UK climate change and security discourse, who at the same time have had a major impact on discussions at the international level, e.g. within the UN.

3 The coding strategy and the use of qualitative data analysis software (see Chapter 3) were well suited to operationalize such a strategy. In a first step, the same heuristic consisting of only a few abstract theoretical concepts was applied to all three cases. These general codes were subsequently filled and supplemented by more empirically enriched sub-codes. After having finished one case, these empirically rich codes were dismissed before turning to the next case, in order not to presuppose any findings. In a next step of analysis, however, the three cases were subsequently reconsidered on the basis of the findings of the previous round of analysis. Thus the important findings and dominant codes from one case were transferred to the other two to search for similarities or to approve the particularity of findings at one level.

4 This means that, for example, the articulation of any actor is not assigned to the UK case because the actor is based within the UK but because the actor refers to the UK in its argument.

References

Austin, John Langshaw. 1975. *How to do things with words*. Oxford: Oxford University Press.

Baldwin, Andrew. 2014. "The political theologies of climate change-induced migration." *Critical Studies on Security* 2 (2): 210–222.

Beckett, Margaret. 2007. "The case for climate security. Lecture by the Foreign Secretary, the Rt. Hon. Margaret Beckett MP, 10/5/2007." Accessed April 30, 2015. https://www.rusi.org/events/past/ref:E464343E93D15A/info:public/infoID:E4643430E3E85A/.

Bloomfield, Steve. 2007. "Climate change leading to global fight for resources." *The Independent*, June 21, 2007.

Boas, Ingrid. 2015. *Climate migration and security. Securitisation as a strategy in climate change politics.* New York: Routledge.

Brauch, Hans G. 2012. "Policy responses to climate change in the Mediterranean and MENA Region during the Anthropocene." In *Climate change, human security and violent conflict*, edited by Jürgen Scheffran, Michael Brzoska, Hans Günter Brauch, Peter M. Link and Janpeter Schilling, 719–794. Berlin/Heidelberg: Springer.

Brown, Oli, Anne Hammill and Robert McLeman. 2007. "Climate change as the 'new' security threat: implications for Africa." *International Affairs* 83 (6): 1141–1154.

Brzoska, Michael. 2009. "The securitization of climate change and the power of conceptions of security." In *Security and Peace* (*Sicherheit und Frieden*) 27 (3): 137–145.

Brzoska, Michael. 2012. "Climate change as a driver of security policy." In *Climate change, human security and violent conflict*, edited by Jürgen Scheffran, Michael Brzoska, Hans Günter Brauch, Peter Michael Link and Janpeter Schilling, 165–184. Berlin/Heidelberg: Springer.

Buckland, Ben. 2007. *A climate of war? Stopping the securitization of climate change.* Geneva: International Peace Bureau.

Buhaug, H., J. Nordkvelle, T. Bernauer, T. Böhmelt, M. Brzoska, J. W. Busby, A. Ciccone, H. Fjelde, E. Gartzke, N. P. Gleditsch, J. A. Goldstone, H. Hegre, H. Holtermann, V. Koubi, J. S. A. Link, P. M. Link, P. Lujala, J. O'Loughlin, C. Raleigh, J. Scheffran, J. Schilling, T. G. Smith, O. M. Theisen, R. S. J. Tol, H. Urdal and N. von Uexkull. 2014. "One effect to rule them all? A comment on climate and conflict." *Climatic Change* 127 (3–4): 391–397.

Buzan, Barry, Ole Wæver and Jap De Wilde. 1998. *Security: a new framework for analysis.* Boulder, CO: Lynne Rienner Publishers.

Climateandsecurity.org, The Center for Climate and Security. 2015. G7: "Climate change, security and fragility as a foreign policy priority." Accessed May 25, 2015. http://climateandsecurity.org/2015/04/17/g7-climate-change-security-and-fragility-as-a-foreign-policy-priority/.

Corry, Olaf. 2014. "From defense to resilience: environmental security beyond neoliberalism." *International Political Sociology* 8 (3): 256–274.

Dalby, Simon. 2015. "Climate geopolitics: securing the global economy." *International Politics*, published Online March 20, 2015.

Davenport, Corel. 2014. "Pentagon signals security risks of climate change." *New York Times*, October 13, 2014, A14.

Detraz, Nicole and Michele M. Betsill. 2009. "Climate change and environmental security: for whom the discourse shifts." *International Studies Perspectives* 10 (3): 303–320.

Dyer, Gwynne. 2009. *Climate wars: how peak oil and the climate crisis will change Canada (and our lives).* Toronto: Vintage Canada.

Fears, Darryl. 2011. "Study sees a link between El Nino and civil unrest." *Washington Post*, August 30, 2011.

Fetzek, Shiloh and Jeffrey Mazo. 2014. "Climate, scarcity and conflict." *Survival* 56 (5): 143–170.

Fleming, James R. 2007. "The climate engineers." *The Wilson Quarterly*: 46–60.

Flyvbjerg, Bent. 2001. *Making social science matter: why social inquiry fails and how it can succeed again.* Cambridge: Cambridge University Press.

Glynos, Jason and David R. Howarth. 2007. *Logics of critical explanation in social and political theory.* London: Routledge.

Hajer, Maarten A. 1995. *The politics of environmental discourse: ecological modernization and the policy process.* Oxford: Clarendon Press.

Hartmann, Betsy. 2014. "Converging on disaster: climate security and the Malthusian anticipatory regime for Africa." *Geopolitics* 19 (4): 757–783.

Jackson, Emily Lindsay. 2010. "Imagining security in the National Security Strategy of the United Kingdom." *In-Spire: Journal of Law, Politics and Societies* 5 (1): 24–39.

Ki-moon, Ban. 2009. *Opening remarks to the United Nations Climate Change Summit Plenary.* New York: United Nations.

Laclau, Ernesto and Chantal Mouffe. 1985. *Hegemony and socialist strategy: towards a radical democratic politics.* London: Verso Books.

Lamothe, Dan. 2014. "Climate change threatens national security, Pentagon says." *Washington Post Online*, October 13, 2014. Accessed April 30, 2015. www.washingtonpost.com/news/checkpoint/wp/2014/10/13/climate-change-threatens-national-security-pentagon-says/.

McDonald, Matt. 2008. "Securitization and the construction of security." *European Journal of International Relations* 14 (4): 563–587.

Methmann, Chris and Angela Oels. 2015. "From 'fearing' to 'empowering' climate refugees: governing climate-induced migration in the name of resilience." *Security Dialogue* 46 (1): 51–68.

Methmann, Chris P. and Delf Rothe. 2013. "Apocalypse Now! From exceptional rhetoric to risk management in global climate politics." In *Interpretive approaches to global climate governance: (De-)constructing the greenhouse*, edited by Chris P. Methmann, Delf Rothe and Benjamin Stephan, 105–121. London: Routledge.

Ministry of Defence. 2010. *Defence in a changing climate*, London: Sustainable Development and Continuity Division. Crown Copyright 04/10.

Oels, Angela and Michael Brzoska. 2012. "From 'securitization' of climate change to 'climatization' of the security field: comparing three theoretical perspectives." In *Climate change, human security and violent conflict*, edited by Jürgen Scheffran, Michael Brzoska, Hans Günter Brauch, Peter Michael Link and Janpeter Schilling, 185–206. Berlin/Heidelberg: Springer.

Pentagon, US Ministry of Defense. 2014. *Climate change adaptation roadmap.* Science and Technology Directorate, US Department of Defense, Alexandria.

Reiber, Tatjana and Fariborz Zelli. 2011. "Klimawandel in der Perspektive verschiedener Sicherheitsdiskurse: Drei Interpretationen und ihre Folgen für die Politik." In *Politik im Klimawandel: keine Macht für gerechte Lösungen?*, edited by Susanne S. Schüttemeyer, 215–238. Baden-Baden: Nomos.

Searle, John R. 1969. *Speech acts: an essay in the philosophy of language.* Cambridge: CUP.

Smith, Paul J. 2007. "Climate change, mass migration and the military response." *Orbis* 51 (3): 617–633.

Solana, Javier. 2008. *Climate change and international security*: Paper from the High Representative and the European Commission to the European Council. European Commission.

Stake, Robert E. 1994. "Case studies." In *Handbook of qualitative research*, edited by Norman K. Denzin and Yvonna S Lincoln, 236–247. Thousand Oaks, CA: Sage.

Theisen, O. M., H. Holtermann and H. Buhaug. 2012. "Climate wars? Assessing the claim that drought breeds conflict." *International Security* 36 (3): 79–106.

Trombetta, Maria J. 2008. "Environmental security and climate change: analysing the discourse." *Cambridge Review of International Affairs* 21 (4): 585–602.

UNGA, United Nations General Assembly. 2009. Sixty-third session, 85th plenary meeting Wednesday, June 3, 2009.

UNSC, United Nations Security Council. 2007a. Sixty-second year, 5663rd meeting, Tuesday, April 17, 2007.

UNSC, United Nations Security Council. 2007b. Sixty-second year, 5663rd meeting, Tuesday, April 17, 2007, Resumption.

UNSC, United Nations Security Council. 2011a. Sixty-sixth year, 6587th meeting, Wednesday, July 20, 2011.

UNSC, United Nations Security Council. 2011b. Sixty-sixth year, 6587th meeting, Wednesday, July 20, 2011, Resumption.

von Lucke, Franziskus, Zehra Wellmann and Thomas Diez. 2014. "What's at stake in securitising climate change? Towards a differentiated approach." *Geopolitics* 19 (4): 857–884.

Wæver, Ole. 1995. "Securitization and desecuritization." In *On security*, edited by Ronnie D. Lipschutz, 46–86. New York: Columbia University Press.

2 Out of school

From Copenhagen through Paris into the world

In search for an analytical approach to study the twofold puzzle identified in the previous chapter, it is necessary to have a closer look at the securitization framework of the Copenhagen School. On the basis of the critique of the Copenhagen School concept in the literature I then introduce the sociological variant of securitization by the so-called Paris School. The chapter finishes with a critique of the school thinking in securitization theory and argues for a revision of the original securitization framework from within.

Widening and broadening the security concept

The core idea of the securitization approach cannot be understood without considering the historical and discursive context of its formulation in the early 1990s, when the collapse of the Soviet Union fuelled attempts to rethink security policy. Throughout the block confrontation of the Cold War, international politics as well as IR theories had been dominated by a narrow, military-centric understanding of security. Following this classical understanding, the nation state uses its monopoly of force to provide for inner security by pacifying social relations as well as for external security by protecting its population and territory from external military threats. As early as the 1970s, however, single voices challenged this narrow understanding of national security under the impression of the global economic crises and the growing awareness of global environmental change (Wæver 1995: Fn. 3). While these voices did not produce much resonance under the omnipresent total threat of nuclear war, the end of the block confrontation finally revealed the historical contingency of the security concept (Stripple 2005: 41–42). The historical juncture brought grist to the mills of those calling for a *broadening* of the security concept beyond the realm of military security as well as a *deepening* of the concept by considering referent objects other than the nation state (Buzan and Hansen 2009: 187). Against this backdrop scholars began challenging the notion of a materialist ontology of security – the idea of security as something given that can be observed by scientists (Stripple 2005: 49).

In his seminal book *Writing security* the poststructuralist scholar David Campbell (1992) outlined the contours of a discursive reading of security based on the idea that security threats are not an objective category but a matter of discursive construction (Campbell 1992: 2). Security, according to Campbell, is a discursive strategy to constitute the dangerous 'radical other' and the self in processes of 'othering' (Campbell 1992: 70–71). Michael Dillon, another early proponent of poststructuralism in IR, argued that security "is not a noun that names something, it is a principle of formation that does things" (Dillon 1996: 16). Such radical calls for a redefinition were heavily attacked – as kind of a defense reflex – by classical thinkers of security. A good example is neorealist Stephan Walt who warned that "[...] by this logic, issues such as pollution, disease, child abuse, or economic recessions could all be viewed as threats to 'security'. Defining the field in this way would destroy its intellectual coherence [...]" (Walt 1991: 213).

The securitization approach of the Copenhagen School[1]

It was against the backdrop of this debate on a redefinition of security that several members of the Copenhagen Peace Research Institute, such as Ole Wæver, Barry Buzan and Jaap de Wilde (Buzan *et al.* 1998), later summarized under the common label Copenhagen School (see McSweeney 1996), developed their framework of securitization. Instead of yet another definition of security, they provided a coherent and slim conceptual framework to study the processes by which security practices had been extended to different sectors – from the political, economic, and societal to the environmental sector. The securitization framework conducted a balancing act between two opposing positions: the 'wideners' call for an extension of security to fields other than military security, and the 'traditionalists' warn against the risk of a conceptual dilution when security could mean anything. Against the backdrop of these arguments, the Copenhagen School sought to develop a broad concept of security that is at the same time coherent (Buzan *et al.* 1998: 4; Wæver 2009: 24). This concept could be found in an understanding of security as a conventionalized speech act (Wæver 1989).

Security as speech act

As mentioned above, the theoretical background of the Copenhagen School's approach of securitization is speech act theory. Speech act theory, which gave birth to pragmatism as a linguistic sub-discipline, was developed by John L. Austin and John Searle. They argued that language does not consist solely of propositions, i.e. statements that carry certain content and can be judged as true or false, but also of performatives. The latter are utterances through which specific actions are performed – or speech acts (Austin 1975; Searle 1969). Good examples of actions performed through speech are commanding, requesting, or asking.[2] Speech acts follow a conventionalized script to which the speaker has to stick and many are bound to the authorized position of certain speakers; 'I now pronounce you husband and wife' in the act of sealing a marriage is a good case in point.

Linguistic pragmatism, then, understands speech acts as utterances that can be distinguished into three different parts: first, the locutionary act, referring to the physical practice of the uttering itself; second, the illocutionary act, the performance of a specific action on the basis of social convention (marriage, order, argumentative attack etc.); and, third, the perlocution, accounting for the speech act's effect on the audience (e.g. the fulfillment of a request) (Austin 1975: 102). The Copenhagen School takes up this idea of the performativity of language and transfers it to the concept of security. A security argument, hence, does not simply describe a certain issue, such as, for example, climate change. Rather, security is to be seen as a conventionalized speech act that constitutes a security problem by naming it, i.e. by way of its illocutionary force (Buzan *et al.* 1998: 26). Drawing on speech act theory, the Copenhagen School develops a framework of securitization, which consists of a few basic concepts. The first one is the *securitizing actor*: "A securitizing actor is someone, or a group, who performs the security speech-act" (Buzan *et al.* 1998: 40). Just as with other conventionalized speech acts, such as a judgment in court or the contraction of marriage through a registrar, a securitizing move can only succeed if uttered from actors in authoritative positions. Securitizing actors must have the authority as well as legitimacy to be heard and believed. Moreover, they must substantiate that they have the capacity to adopt measures to deal with the problem. These assumptions limit the possible candidates for securitizing actors to governments and other leading politicians or bureaucrats, or in cases of highly specialized issue areas, scientific experts or technocrats (Buzan *et al.* 1998: 40).

The second element of the Copenhagen School framework is the *securitizing move* – the performative act constituting a security problem. Securitizing moves draw on a specific script – a 'grammar of security' – that is socially conventionalized and hence understood by the audience. This script follows from a realist notion of security and involves the identification of an existential threat to a referent object (a country, political community etc.), a sense of urgency and a call for exceptional measures to avert the threat (Wæver 1995: 49).[3] The last component of the securitization approach is the *relevant audience* and the perlocutionary effect of the utterance. If accepted by a relevant audience, the securitizing move constructs a sense of urgency and legitimizes an adoption of a range of exceptional measures[4] to deal with the problem that would not be accepted for 'regular' political problems (Wæver 1995: 55). The intentional and tolerated 'breaking of rules' is thus the decisive criterion in identifying the perlocutionary effect of successful securitization (Buzan *et al.* 1998: 24). Yet, according to Buzan *et al.*, the securitizing actor alone never constructs a security problem. Rather, a successful securitization rests upon a negotiation process between speaker and audience. In concrete terms, for the Copenhagen School this implies that securitizing moves can fail if a relevant audience does not accept them.

While, according to the Copenhagen School, nothing is a security problem by nature and everything could potentially be securitized, the success of a securitizing move is nevertheless dependent on the characteristics of the constructed threat as well as the referent object. On the one hand, the characteristics of the constructed security problem exert an influence: the more this issue is generally associated with notions of threat and danger, the likelier the securitization is to succeed (Buzan *et al.* 1998: 33). On the other hand, the political scale of the referent object can decide the success or failure of a securitization. The approach of the Copenhagen School suggests that successful securitization in most cases will be restricted to the mid-range political scale, i.e. the level between the single individual and the (international) system level. As "referent objects must establish security legitimacy in terms of a claim to survival" (Buzan *et al.* 1998: 39), smaller units such as the individual are less successful in securitizing moves. Single individuals simply lack the political weight to bring about public interest in their security. According to Buzan *et al.*, systemic units such as the international community or humanity are seldom successfully addressed in securitizing moves, due to their high degree of abstraction as well as the absence of potential enemies. Political communities such as nation states, on the contrary, allow for a demarcation from the outside/other as well as social cohesion through shared values and interpretations, guaranteeing a strong identification with the referent object (Buzan *et al.* 1998: 36–37).

In a 2009 article, Buzan and Wæver (Buzan and Wæver 2009) partly revise their reflections on the referent object of securitizations. Concretely, they present a possibility of how different smaller securitizations might be linked to more comprehensive security discourses at the transnational or even global level. Such higher-level discourses could be understood as macro-securitizations

> [...] that speak to referent objects higher than those at the middle level (for example, 'universal' religions or political ideologies; one or more of the primary institutions of international society) and which aim to incorporate and coordinate multiple lower level securitisations.
> (Buzan and Wæver 2009: 257)

The global discourse on terror or the threat of climate change, which affect the terrorism or climate change discourses within different countries, are good examples of such macro-securitizations. Thus, the macro-securitization concept is an important extension of the theory with respect to the case under investigation in this book.

Critique of the speech act model in the literature

The securitization approach of the Copenhagen School has attracted a great deal of attention from IR and security studies. The concept has prompted a variety of empirical investigations of different topics that have recently become seen as security problems.[5] No less important are works that have addressed the theoretic deficits of the securitization. Such theoretic criticism is almost exclusively voiced from positions that sympathize with the approach in general but have the aim to further advance it (see Buzan and Hansen 2009: 216).

The most frequent point raised by critics is the *narrowness* of the securitization concept (e.g. McDonald 2008: 567; Stritzel 2007: 359). According to critics, this narrowness results from a biased concentration on securitizing speech acts and the ignoring of the discursive and material *context* of these speech acts. Critics such as Thierry Balzacq have ascribed this 'internalist perspective' to the language philosophical foundation of the Copenhagen School: "The assumption of a speech act approach ultimately reduces security to a conventional procedure such as marriage or betting [...]" (Balzacq 2005: 172). Different scholars have argued that the success of a speech act rests upon its specific context and cannot be reduced to the content of the utterance (Stritzel 2007: 369). On the one hand, this refers to the material context of the speaker's positions (Balzacq 2005) including the speaker's social and symbolic capital and socio-economic power relations. On the other hand, this includes the ideational context – the already existing discourses in the respective fields, which enable certain utterances in the first place.

A point related to the internalist perspective of the Copenhagen School is the way it conceptualizes the author and the relative audience. On the one hand, given the criteria that securitizing actors have to meet according to Buzan *et al.*, there is a clear focus on national decision makers. This focus is problematic as it blurs the fact that discourses are shaped by many competing and often oppositional voices (e.g. Williams 2003: 527–528). As the Copenhagen School does not account for such oppositional voices, Kenneth Booth, for example, argues that "securitization studies therefore suffer from being elitist" (Booth 2007: 166). On the other hand, the passive role of the audience in the framework of the Copenhagen School has been criticized. The Copenhagen School reduces the audience to the decision between success and failure of a securitizing move. According to Balzacq this can be traced back to the theoretical assumptions of speech act theory and its distinction between illocution – the performed act – and perlocution – the effects of the act on the audience (Balzacq 2010: 5). According to this critique, the effects and implications of securitization are causally ascribed to the illocutionary power of language. If the perlocution, however, were regarded as part of the securitization process itself, a much more (inter)active role of the audience would become visible. This inclusion would also allow for the consideration of a whole range or continuum of possible effects rather than solely for the binary outcome of success or failure (Vuori 2008: 71).

By focusing on the illocutionary force of the speech act, a securitization process furthermore takes place at *one* particular moment – namely between the articulation of the actor and the acceptance/rejection of the audience. However, empirical investigations have shown that identifying the exact moment of securitization is problematic (Stritzel 2007: 366).[6] As a solution, different authors have suggested that securitization should be understood as a continuous process in which a security problem is constructed inter-subjectively through the practices of different actors (Balzacq 2010: 3; McDonald 2008: 576).

Finally, critics have also pointed to the normative bias that characterizes the Copenhagen School approach. As securitization comes along with a depoliticization or de-democratization of issues, the approach advocates a desecuritization program (Aradau 2004). On the contrary, some have argued that securitization processes could actually have positive effects such as raising political awareness for particular issues (Floyd 2007: 43–44). This normative bias towards desecuritization, hence runs contrary to the initial attempt to provide a neutral analytical tool to study the expansion of the security field (Taureck 2006: 58).

Towards a sociological theory of securitization ... and beyond

The critique of the blind-spots resulting from the language–philosophical roots of the Copenhagen approach (Stritzel 2007: 359) has spurred the formulation of an alternative, sociological variant of securitization theory usually associated with the so-called Paris School (see e.g. Balzacq 2010; Balzacq and Guzzini 2015: 97–101). In search of theoretical alternatives that could help overcome the conceptual blind spots of the securitization approach in the case of climate change outlined in the introduction, the Paris School should hence be the first option. In short the Paris School develops a more context-sensitive, processual notion of securitization, which understands securitization as an inter-subjective process (Balzacq 2010: 3). It broadens the analytical perspective beyond speech acts as lingual practices to include a broader range of performative acts and practices in the production of security threats. The main concepts in this approach are the concepts of the *field* and the *habitus* by Pierre Bourdieu on the one hand and Foucault's notion of the *dispositif* on the other. Following Bourdieu, habitus refers to a set of dispositions[7] that characterize either the position of the securitizing actor or a whole social field.[8] The habitus of a speaker position builds the basis of its social and cultural capital, e.g. the social credibility of scientists, and hence of the ability to participate in processes of securitization. The habitus of the social field, on the contrary, helps in understanding the routinization of certain practices in a given field – such as practices of surveillance in the security field (Balzacq 2010: 2).

Drawing on the concept of the field should allow for the accounting of the socio-cultural embeddedness of texts (speech acts). Along the same lines, the notion of the *dispositif* should help overcome the fixation on language in the securitization approach. Michel Foucault established the notion of the dispositif as an extension to his concept of discourse. The dispositif covers not only linguistic practices but also the interaction of institutions, practices, scientific assumptions, policies etc. – "in short, the said as much as the unsaid" (Foucault 1972: 194) in a given social field. By understanding security as a *dispositif* one can study how very heterogeneous practices, techniques, articulations etc. are linked together in the production of security. It is this heterogeneity of different elements that constructs different issues as security problems.

The sociological model of the Paris School holds that security problems are not constructed through the 'magical power' (performativity) of language. Rather, an issue becomes securitized because it is institutionalized within the security field through repetition and habituation in practices of security actors (Balzacq 2010: 13). An illustrative example of a sociological, practice-based understanding of securitization is provided by Didier Bigo in his study of the securitization of the EU's immigration policy (Bigo 2002). This securitization, according to Bigo, is not based on "a new and more efficient rhetoric convincing the population of a danger, or successful 'speech acts' performed by actors coming from the state or from the society" (Bigo 2002: 65). Rather, Bigo reveals, a securitization of immigration to the EU primarily took place through the routinized and ongoing practices of surveillance and control of security professionals. The practices of actors such as police forces, the military, private security companies, custom officers, and also humanitarian actors – for Bigo the 'managers of unease' – constitute a security field in which particular problems such as migration are translated into security threats (Bigo 2002: 63–65). "Securitization of the immigrant as a risk [...] is structured by the habitus of the security professionals and their new interests not only in the foreigner but in the 'immigrant'" (Bigo 2002: 65).

The approach of the Paris School presents an important and sophisticated revision of the securitization concept, which provides valuable insights for the case under investigation in this book – such as opening the approach to non-linguistic practices, the processual understanding of securitization or the suggestion of studying the relation between security and politics in degrees, not in dichotomic terms (Balzacq 2015: 108). For example, it sheds light on the important role of experts and their – often unnoticed and mundane – practices such as mapping or computer modeling in the securitization of climate change (see Chapters 5 and 6). That being said, the Paris School approach is itself not without problems. Its orientation towards sociology means that some of the initial disciplinary debates in IR, which gave birth to the securitization approach, disappear from view. This holds true for debates on Schmittian politics (of exception) and the construction of antagonist identities through processes of 'othering,' which, even after the Cold War, play a major role in international politics. Security theorizing is dismissed in favor of neo-empiricist analyses of security practices at the micro-level. "This allows the sociological 'theory' to become an unstructured mapping of actors' perceptions/predispositions and their exchanges of communication – with no theoretical focal idea, in particular no organizing theoretical tool at the collective, political level" (Wæver 2015: 125). The Paris School's approach, furthermore, dismisses the category of discourse as an ontological and epistemological category in its own right in order to focus on the interaction of diverse security practitioners. As Hansen argues, within the promoted externalist approach, language and discourse are only epiphenomenal – they do not construct security problems but at best frame external and existing threats in a particular way (Hansen 2011: 359).[9] This risks undoing the major contribution of the Copenhagen School to the literature of security, namely, the particular reading of security that defines security problems as neither objective nor subjective but rather as discursive. A related problem is that the understanding of securitization as *intersubjective* process contradicts other conceptual elements of the Paris School approach borrowed from Michel Foucault, as, for Foucault, a discourse or dispositif has to be considered as a distinct ontological phenomenon independent of and irreducible to perceptions and (symbolic) interactions of individual subjects. In addition, "Balzacq's claim that Buzan and Wæver take an internalist position [...] reinstalls the materialism–idealism distinction that poststructuralists have laboured to deconstruct" (Hansen 2011: 360).

An even bigger problem is the fashion in which the Paris School counterposes the Copenhagen School through a dichotomous distinction between a philosophical/poststructuralist and a sociological version of securitization (Balzacq and Guzzini 2015: 103). This binary framing is an artificial construct which not only is analytically unrewarding but moreover unnecessarily narrows our conceptual perspective. "If one comes to securitization theory through poststructuralism, [...] it is not clear that Buzan and Wæver face the problems that Balzacq and Stritzel attribute to them" (Hansen 2011: 359). And if one furthermore takes a closer look at the different trajectories of thought which back the two approaches, one identifies many overlaps between thinkers such as Searle and Derrida on the one hand and Bourdieu and Foucault on the other. By way of conclusion: the Paris School risks undoing some of the benefits that a discursive

Table 2.1 Conceptual merits and problems of the Paris School's approach

Merits	*Problems*
Securitization as an inter-subjective process	Too narrow a focus on actors and their practices at the micro-level; ignorance of the collective, political level
Inclusion of linguistic and non-linguistic practices	Dismissal of discourse as ontological and epistemological concept
Openness to different political implications; rejection of exceptionality as analytical yardstick	None
Sociological perspective: habitus/field; dispositif	Antagonistic opposition to philosophical/poststructuralist version of the CS → unnecessary and not reasonable

reading of security has brought to IR. Table 2.1 summarizes the merits and problems of the Paris School approach compared to the original securitization framework.

The school logic that has informed the debate on securitization and the underlying binary juxtaposition of a sociological and poststructuralist version of securitization obscures more than it clarifies. And it structures and divides the field in a way that prevents fruitful synergies between different theories and modes of thought being discovered. In the following theoretical discussion I show that, through a consequently poststructuralist re-reading of securitization, many of the constructed oppositions dissolve. In other words, the conceptual shortcomings of the Copenhagen School leveled in the literature do not stem from the adoption of discourse theory but rather from contradictions of the poststructuralist elements of the theory with other elements, such as speech act theory or a Schmittian concept of the exception. This means that it is more promising to revise the framework *from within* rather than to attack the Copenhagen School approach from the outside by presenting an oppositional approach. Rather than juxtaposing the Copenhagen School approach with theoretical alternatives, as is done by the Paris School, I see the best possibility for advancing securitization in a *revision from within*.

Notes

1 Short passages of this section have already been published elsewhere (see Rothe 2012).

2 Later Searle rejected the distinction between propositional utterances that state something and performative utterances that do something. Instead, he argued that every utterance comprises a propositional and a performative dimension (Searle 1969).

3 However, Buzan *et al.* also point to the fact that different sectors are characterized by a specified discursive vocabulary which can differ from a realist grammar of security (Buzan *et al.* 1998: 33).
4 The Copenhagen School acknowledges the problem of defining exceptional measures. As a way out they circumvent giving any positive definition but instead define exceptionality as a 'breaking' of established rules. What counts as exceptional hence always depends on the given rules and norms in a political community (Buzan *et al.* 1998: 26).
5 The literature includes studies of the securitization of international migration and terrorism (e.g. Huysmans 2000), pandemics like bird flu or HIV/Aids (Elbe 2008; Youde 2008), or piracy (Tsvetkova 2009).
6 First of all, it is not easy to identify the relevant audience unless securitizing actors explicitly address a particular group. In the end, the identification of the audience in most cases will depend on a subjective judgment by the researcher. But even if one convincingly identifies the relevant audience, it is hard to measure the success of the securitizing move, as people will seldom explicitly express their acceptance or rejection. Furthermore, the only measurable criterion for a successful securitization is the adoption of exceptional measures subsequent to a securitizing move. From this observation one could ex-post infer the successful acceptance of a securitizing move through the audience.
7 Dispositions are understood as characteristics such as habits, conventions, lifestyles, taste etc., which structure the perception and behavior of actors and at the same time represent the basis for their social or cultural capital (Bourdieu 1997). It is important to note, however, that these dispositions are not a feature of the individual person, but of the subject position (as an academic, a politician, an athlete etc.) he or she adopts.
8 According to Bourdieu, the field is an analytical category to study societal interaction. The space of the social is permeated by a variety of specialized fields which each have their own functional logics, norms, rules and institutions. The totality of these rules is rooted in the habitus of the subject positions in a field. As the behavior of these subjects is influenced by these internalized habitus, they at the same time reproduce these constitutive rules (Bourdieu 1997).
9 Balzacq himself indicates the secondary, epiphenomenal role of language writing that "language does not construct reality; at best, it shapes our perception of it" (Balzacq 2010: 12).

References

Aradau, Claudia. 2004. "Security and the democratic scene: desecuritization and emancipation." *Journal of International Relations and Development* 7: 388–413.

Austin, John Langshaw. 1975. *How to do things with words.* Oxford: Oxford University Press.

Balzacq, Thierry. 2005. "The three faces of securitization: Political agency, audience and context." *European Journal of International Relations* 11 (2): 171–201.

Balzacq, Thierry. 2010. "A Theory of securitization: origins, core assumptions, and variants." In *Securitization theory: how security problems emerge and dissolve*, edited by Thierry Balzacq, 1–30. London, New York: Routledge.

Balzacq, Thierry. 2015. "The 'essence' of securitization: theory, ideal type, and a sociological science of security." *International Relations* 29 (1): 103–113.

Balzacq, Thierry and Stefano Guzzini. 2015. "Introduction: 'What kind of theory – if any – is securitization?'." *International Relations* 29 (1): 97–102.

Bigo, Didier. 2002. "Security and immigration: toward a critique of the governmentality of unease." *Alternatives: global, local, political* 27 (1 suppl): 63–92.
Booth, Ken. 2007. *Theory of world security, Cambridge studies in international relations, 105*. Cambridge: Cambridge University Press.
Bourdieu, Pierre. 1997. *Der Tote packt den Lebenden, Schriften zu Politik und Kultur, 2.* Hamburg: VSA-Verl.
Buzan, Barry and Lene Hansen. 2009. *The evolution of international security studies.* Cambridge: Cambridge University Press.
Buzan, Barry and Ole Wæver 2009. "Macrosecuritisation and security constellations: reconsidering scale in securitisation theory." *Review of International Studies* 35 (02): 253–276.
Buzan, Barry, Ole Wæver and Jap De Wilde. 1998. *Security: a new framework for analysis.* Boulder, CO: Lynne Rienner Publishers.
Campbell, David. 1992. *Writing security. US foreign policy and the politics of identity.* Minneapolis: University of Minnesota Press.
Dillon, Michael. 1996. Politics of security. Towards a continental philosophy of continental thought. London: Routledge.
Elbe, Stefan. 2008. "Risking lives: AIDS, security and three concepts of risk." *Security Dialogue* 39 (2–3): 177–198.
Floyd, Rita. 2007. "Towards a consequentialist evaluation of security: bringing together the Copenhagen and the Welsh Schools of security studies." *Review of International Studies* 33 (2): 327–350.
Foucault, Michel. 1972. "The confession of the flesh." In *Power/knowledge: selected interviews and other writings*, edited by Colin Gordon, 194–228. Brighton, Sussex: Harvester Press.
Hansen, Lene. 2011. "The politics of securitization and the Muhammad cartoon crisis: A post-structuralist perspective." *Security Dialogue* 42 (4–5): 357–369.
Huysmans, Jef. 2000. "The European Union and the securitization of migration." *Journal of Common Market Studies* 38 (5): 751–777.
McDonald, Matt. 2008. "Securitization and the construction of security." *European Journal of International Relations* 14 (4): 563–587.
McSweeney, Bill. 1996. "Identity and security: Buzan and the Copenhagen School." *Review of International Studies* 22 (1): 88–91.
Rothe, Delf. 2012. "Security as a weapon: how cataclysm discourses frame international climate negotiations." In *Climate change, human security and violent conflict*, edited by Jürgen Scheffran, Michael Brzoska, Hans Günter Brauch, Peter M. Link and Janpeter Schilling, 243–258. Berlin/Heidelberg: Springer.
Searle, John R. 1969. *Speech acts: An essay in the philosophy of language.* Cambridge: CUP.
Stripple, Johannes. 2005. *Climate change after the international: rethinking security, territory and authority.* Lund: Department of Political Science, Lund University.
Stritzel, Holger. 2007. "Towards a theory of securitization: Copenhagen and beyond." *European Journal of International Relations* 13 (3): 357–383.
Taureck, Rita. 2006. "Securitization theory and securitization studies." *Journal of International Relations and Development* 9 (1): 53–61.
Tsvetkova, Bilyana. 2009. "Securitizing piracy off the coast of Somalia." *Central European Journal of International and Security Studies* 3 (1): 44–63.
Vuori, J. A. 2008. "Illocutionary logic and strands of securitization: applying the theory of securitization to the study of non-democratic political orders." *European Journal of International Relations* 14 (1): 65–99.

Wæver, Ole. 1989. "Security, the speech act: analysing the politics of a word." In *Working Paper 19.* Copenhagen: Copenhagen Peace Research Institute.

Wæver, Ole. 1995. "Securitization and desecuritization." In *On security*, edited by Ronnie D. Lipschutz, 46–86. New York: Columbia University Press.

Wæver, Ole. 2009. "What exactly makes a continuous existential threat existential – and how is it discontinued?" In *Existential threats and civil-security relations*, edited by Oren Barak and Gabriel Sheffer, 19–36. Plymouth, MA: Lexington Books.

Wæver, Ole. 2015. "The theory act: responsibility and exactitude as seen from securitization." *International Relations* 29 (1): 121–127.

Walt, Stephen M. 1991. "The renaissance of security studies." *International Studies Quarterly* 35 (2): 211–239.

Williams, M. C. 2003. "Words, images, enemies: securitization and international politics." *International Studies Quarterly* 47 (4): 511–531.

Youde, Jeremy. 2008. "Who's afraid of a chicken? Securitization and Avian Flu." *Democracy and Security* 4 (2): 148–169.

3 Securitization deconstructed

Towards a consequent poststructuralist framework

In the last chapter I have outlined the shortcomings that have been identified in both the frameworks of the Copenhagen School and the Paris School on securitization. In particular I have argued against the problematic distinction between a poststructuralist and a sociological account of securitization. In this chapter I thus set out to revise the securitization framework of the Copenhagen School from within in order to excavate the contours of a consequent poststructuralist theory of securitization. Herein, I follow a method of *deconstruction* in a twofold sense. First of all, deconstruction means isolating the different building blocks of the framework and tracing back their conceptual roots. Thereby it becomes possible to polish every module until the framework is based on a more solid and coherent foundation. Second, deconstruction in line with Derrida (Derrida 1976) refers to a more particular method to reveal the existence of a variety of antinomies within the securitization approach.[1] In the chapter I reveal the major antinomies and shortcomings of the Copenhagen School approach that stem from an impossible marriage of theories such as poststructuralism and IR realism (see Taureck 2006: 23–26) while keeping the progressive potential of a discursive reading of security.[2] I will show that, if one takes the assumptions of poststructuralist discourse theory seriously, much of the critique raised in the literature can be dissolved. For each of the critical points identified, I begin with a brief description of the recent critique and debate in the literature and I then propose an alternative reading. Together these revisions constitute the building blocks for a consequently poststructuralist theory of securitization.[3]

From security as speech act to security as discourse

As I have outlined in the previous chapter, the Copenhagen School approach is often associated with a poststructuralist discourse theory (e.g. Balzacq 2010c: 60; Taureck 2006). In fact, while the Copenhagen School draws on some ideas of poststructuralist thinking, its primary theoretical background is speech act theory. As outlined in more detail above, this language–philosophical foundation of securitization theory has attracted a considerable amount of criticism from the literature. Critics have first questioned the narrow focus on language or linguistic practices as a medium of securitization. Second, scholars have criticized the concentration on single acts, i.e. the concrete events of articulation, and argued that securitization is better understood as an ongoing process. Last, there is criticism that the Copenhagen School pays too much attention to the securitizing actor – the speakers of security – while the role of the audience in securitization processes is underrated.

In this section I outline an alternative to the speech act approach to securitization. My starting point is the assumption that the outlined problems can be overcome if one dissolves the unholy conjunction of speech act theory and poststructuralism. By equating discourse with speech acts, critics misinterpret poststructuralism as a form of linguistic constructivism. If one adopts a consequentially poststructuralist perspective on securitization, the latter can however, be understood as a process of discursive struggle including both linguistic and non-linguistic means of articulating security. Poststructuralist discourse theory also addresses the critique of the focus on the single event of the securitizing act and the concentration on the speaker of securitization. Let me use the following sections to sketch out the contours of securitization, in which the speech act model is dismissed in favor of a comprehensive understanding of discourse. Instead of single securitizing moves and their illocutionary force, the concept of discourse coalitions is placed at the center of such a revised model of securitization. To get there it is necessary to dig a little deeper into the conceptual roots of securitization theory and discuss the ontological and epistemological traces of speech act as well as poststructuralist discourse theory.

The ideational and the material

The theoretical premise of speech acts in the Copenhagen School approach translates epistemologically into a focus on linguistic representations of security in texts. Yet in this section I show that this focus on linguistic practices is not justified through recourse to poststructuralist discourse theory. This is because discourse from a poststructuralist perspective is not reduced to the realm of language (Laclau and Mouffe 1985: 107). Discourse here simply refers to a structure of elements whose meaning is constituted through their relations to the other elements in the structure. Discourses are understood as "[...] relational configurations of elements that comprise agents (or subjects), words, actions, and things" (Howarth 2010: 312). This all-encompassing understanding of discourse has its roots in the linguistic structuralism of Ferdinand de Saussure. Let me shortly outline this conceptual trace a little more.

De Saussure understood language as a differential system of signs (De Saussure 1967: 65–70). The relation between a signifier and its referent, according to de Saussure, is 'arbitrary' – i.e. a sign is not motivated by its referent nor vice versa – other than for example in pictographic script or echoism (De Saussure 1967: 73). Rather, the meaning of a linguistic sign emerges through its differential relation with all the other signs in the system of language.[4] Poststructuralism, then, draws on these assumptions and enriches them with an encompassing theory of discourse. While for structuralism the relation between the signifier and the signified is unmotivated, meaning is nevertheless fixed on the basis of a conventionalized code, shared by members of a speech community. Poststructuralism scraps this notion of language as a stable structure, as the only possibility of fixing an indefinite differential system is the existence of a stable center or focal point (Derrida *et al.* 1974: 22–26). Yet in discourse there is nothing like a universal, atemporal sign, which could take up this function as a stable center. There is nothing other than potentially infinite relations of signs. These are structured along the two logics of equivalence and difference (Laclau and Mouffe 1985: 127–134), following from de Saussure's syntagmatic and paradigmatic relations (see fn. 4).[5]

This assumption of a potentially infinite, unfixed relational structure of signs has two implications. On the one hand, if a closure of the system of signs is impossible and if meaning is dependent on differential relations of signs, meaning must be radically contingent (Glynos and Howarth 2007: 109–110). Meaning is hence subject to a continuous negotiation or a discursive struggle in which the meaning of phenomena is fought out. On the other hand, this assumption implies that there is nothing outside of discourse (Derrida *et al.* 1974: 158). In contrast to the assumption of some critics this *does not imply* that a material reality would not exist. Yet no material object or phenomenon has any meaning to us, if it is not embedded in a relational structure. The discursive constitution of meaning is the mere result of the relationality of objects and signs and not of any intentional speech act or linguistic framing. The distinction made by de Saussure between the object of the world (the signified) and its mental representation (the signifier) becomes blurry. A material object can always function as a sign – as it is relationally embedded with other objects – while any practice – lingual or not – can produce discursive effects by intervening in the relational order of things. This radical assumption of a 'material semiotics' (Law 2009) transcends the unholy dichotomy between speech and practice or the discursive and the material that haunts the debate on securitization theory in the literature.

A storm for instance certainly exists independently from language and impacts humans in a direct manner. Its meaning depends on its associative as well as differential relations within a broader discursive context: it could be understood as a divine punishment, as a random tragedy, or as a consequence of climate change.[6] At the same time, the material characteristics, for example the destructive forces of a super-storm such as Hurricane Katrina, have certain discursive effects – for example by dislocating established structures of meaning or by forging new relations between hitherto unrelated things – for example climate change, natural disasters and US national security.

If discourses are relational structures, *articulation* denotes the practices that shape these structures. Yet, in a broad – material semiotic – understanding of discourse, the practice of articulation is not reduced to speech and writing but refers to every practice that relates elements along the logics of difference and equivalence (Laclau and Mouffe 1985: 105). One could say that while speech act theory stresses the pragmatic quality of language, i.e. the ability to act through language, poststructuralism highlights the articulatory character of non-linguistic practices. Discourses "weave together semantic aspects of language and pragmatic aspects of action" (Torfing 2005: 14). All practices, as far as they are meaningful, impact the relational structure of meaning. In line with Wittgenstein's language games (Wittgenstein 2010: 3), Laclau and Mouffe argue that an object's meaning is constituted not only by how we linguistically denote it but also by how and in what context we use it (Laclau 1990: 100–103).[7] From this broad understanding of discourse and practices of articulation, it follows that performativity should not be restricted to linguistic practices, or speech acts. Rather, performativity refers to the ability to perform a certain phenomenon in a particular way. In the case of a marriage, it is indeed a linguistic speech act that performs the phenomenon of a marriage. In the just mentioned case of a storm it might instead be the practices of praying and penance that perform it as a divine punishment; or it might be the computer models of climate scientists that perform the storm as an effect of climate change. In a national preparedness exercise, finally, a storm event might be performed as a catastrophic threat, thereby constituting a sense of urgency and emotive responsive that no speech act could invoke.

With respect to a theory of securitization, this implies that, on the one hand, a restriction to language (as in the Copenhagen School) as a medium of securitization is unsatisfactory. On the other hand, the separation between linguistic and material practices (as in the Paris School) is equally wrong, just as the assumption that a concentration on text would follow from a commitment to poststructuralist discourse theory. In short, security problems are neither simply out there, nor are they constructed solely through speech acts. Rather, they are articulated, or enacted, by a whole range of lingual as well as non-lingual practices.

Securitization as discursive struggle

It has been argued that one should actually understand securitization in the Copenhagen School's framework as a process and that Buzan *et al.* did not claim at all that securitization would take place in one particular act (Guzzini 2011: 331). This is supported by statements by the Copenhagen School that securitization is based on intersubjective processes (Buzan *et al.* 1998: 25–26) and by its empirical analyses that do not focus on single utterances but, for example, study different discursive layers of security (see e.g. Wæver 2002). However, this processual understanding is not mirrored in the ontological concepts that make up the analytical framework laid out by the Copenhagen School – the securitizing actor, the securitizing move and the perlocutionary effect (the adoption of exceptional measures). In the same volume in which Guzzini argued for a processual understanding of the Copenhagen School's approach, Ole Wæver himself stated that "the structure of securitization theory is organized around securitization as *an act*, a productive *moment* [...]" (Wæver 2011: 468, emphasis added).

What one can notice is a contradiction between a general philosophical ontology of securitization, which understands security as a radically contingent discourse, and a theoretical ontology, which squeezes this assumption into an individualistic framework derived from speech act theory.[8] Put more simply, if I would follow a processual understanding of securitization and discourse, why would I commit myself to an analytical framework that focuses on individual speech *acts*?[9] To develop a theoretical framework that is in line with the general philosophical ontology of the Copenhagen School I would hence suggest understanding securitization as an ongoing discursive struggle. While it is right that securitization is about the redistribution of rights and responsibilities between political actors, as Ole Wæver (2015: 122) has recently argued, it is impossible to attribute this redistribution to a particular illocutionary event. It is not possible to identify a certain moment of securitization. Securitization processes are not only "dynamic, reflectivist and processual" (Stritzel 2011: 344) but often contradictory and messy.

This argument is supported by Derrida's critique of Austin and Searle's speech act model in the late 1980s. For Derrida, the problem of speech act theory was that the meaning of a speech act, i.e. the concrete type of action it performs, depends on the existence of an exhaustive context (Derrida 1988: 14). Put more simply, one needs to know the social position of a speaker, the exact circumstances of the utterance and the established social conventions to identify the meaning of a particular act. What represents an order in one context might be a joke in another context, for example if uttered in a theatre play. Nothing could escape the totality of this context, which implies that there is no room for the polyvalence or ambiguity of utterances (Derrida 1988: 16). The failure of speech acts, e.g. due to a misunderstanding, is excluded as an abnormality. Following Derrida, however, the failure of a speech act is not an abnormal incident (Derrida 1988: 16–19). Ambiguity and the radical contingency of meaning is a crucial feature of discourse, as I have already shown in the last section.

Derrida substantiates these claims with his theoretical concept of *iterability* – the principle of the repeatability of signs. As language is based on conventions and a socially agreed code, every meaningful utterance is repeatable outside of its specific context. At the same time, iterability refers to the transformation of meaning through repetition in different contexts – *iter* is derived from *itara* (which means 'different' in Sanskrit) as Derrida emphasizes (Derrida 1988: 7). For Derrida this most importantly implies an "irreducible absence of intention or attendance to the performative utterance" (Derrida 1988: 18–19). In other words, a perfomative action is not reducible to the intention of a speaker and no speaker holds control over the performative effects of his or her articulation. This argument poses a serious challenge to the Copenhagen School's approach, which holds a securitizing move as a purposeful act to take an issue out of regular political procedure.

The group of the Copenhagen School accounted for Derrida's critique in two ways (Taureck 2006: 9–12). First, it accounted for the possibility of failing as an intrinsic feature of any securitizing move. The fact that a securitizing move succeeded once does not guarantee a renewed success when repeated. The speaker position, the nature of the referent object and the conventionalized script of the securitizing move – in short, the text and the context – altogether do not determine but can only facilitate a securitization of an issue (Stritzel 2007: 361; Taureck 2006: 11). Second, the Copenhagen School took Derrida's assumption, that one could not infer from the content of a speech act to the intention of the speaker, as a justification for excluding context. Following this reading, the text of a speech act constitutes its context and not vice versa. This explains the "centrality of studying in a text, how it produces meaning, rather than relating it to a 'context'" (Wæver 2004) in the approach of the Copenhagen School. Iterability, following this reading, means that the intrinsic characteristics of text/discourse alone decide a securitization, making an analysis of context unnecessary (Balzacq 2010c: 21). As mentioned earlier, such a marginalization of context has been one of the major points of criticism in the literature.

I would argue, however, that both the Copenhagen School with its exclusion of context, and the critique by Balzacq and others (e.g. Balzacq 2010c; Stritzel 2011) do Derrida an injustice. The latter had precisely not claimed that the meaning of an utterance would be founded in itself: "To reject the belief in 'intentions' or 'inner pictures' behind the utterances, however, does not amount to endorsing the belief in any simple adequation of the utterance to itself" (Derrida 1988: 72). What is more, he never did argue that an utterance could be transferred to another context "with little quibble" (Balzacq 2010c: 21). The idea of iterability at the same time implies identity but also difference: a sign is always dependent on the specific 'deformations' inflicted by a concrete context. By stating that "there is nothing outside of the text" (Derrida 1976: 158) Derrida did not only refer to text as a concrete language event but also to an abstract textuality that is comparable to the poststructuralist definition of discourse.

Taken together, Derrida's arguments speak to anything but a narrowing of securitization to the text of a single speech act. His arguments do not speak against but rather for an analysis of context. The important point is how to conceptualize this context. Derrida rejected the notion of an exhaustive, fully constituted context. For him context is an indefinite and always incomplete symbolic structure (Derrida 1976: 18) – the field of the discursive as Laclau and Mouffe have called it (Laclau and Mouffe 1985: 111). The centerpiece of Derrida's critique is not the opposition between text and context but the radical contingency of meaning (Howarth 2000: 42). A broader discursive context at the same time 'deforms' the meaning of an articulation and is itself constantly changed through repeated articulations. Discursive meaning is not produced through the event of a speech act but emerges between them – as a pattern of repeated articulations. "In the first instance, performativity must be understood not as a singular or deliberate act, but, rather, as the reiterative and citational practice by which discourse produces the effects that it names" (Butler 1993: 2). Any attempt to break down these citational chains of performativity to identify a certain illocutionary event necessarily is an analytical reduction. No actor in the world, however powerful she might be, could constitute climate change as a security issue through her articulation. Only through repetition and rearticulation of arguments that link climate change to security could a genuine climate security discourse develop. And the repeated articulation of statements such as 'climate change represents a global security issue' at the same time affects the context of articulations itself – for example the dominant security concepts could slowly change, e.g. from a national to a global reference point of security.

From speaker–audience dichotomy to discourse coalitions

The analytical framework of the Copenhagen School draws on a theoretical ontology derived from speech act theory revolving around the securitizing actor and the securitizing move, which contradicts the general notion of security as a discourse. In the framework of the Copenhagen School, the single speaker of security takes up the central position: "by saying security a state representative declares an emergency condition" (Buzan *et al.* 1998: 21); "something is a security problem when the elites declare it to be so" (Wæver 1995: 54). As these quotes demonstrate, the securitization of an issue follows a specific calculus of the securitizing actor. The meaning of security and discursive changes are hence bound to the intention of the authorized actor. The constitution of the actor's intentions and his or her identity as the basis of intentionality, however, lie beyond the research interest of the framework. Wæver and Buzan justified this bracketing through recourse to Judith Butler's reading of speech act theory, which argued that subjectivity is produced through the performativity of the expressions itself. The actor obtains a temporary status as authorized speaker or subject only by rearticulating particular discursive storylines and producing itself as the 'author' of a statement (Butler 1997).

However, I would argue that a poststructuralist understanding of subjectivity (including Butler's) does not justify the bracketing of the constitution of identity (see Hansen 2011: 360). In this sense the Copenhagen School draws the wrong conclusion from Butler's argument. The most important insight one can derive from Butler and other poststructuralists is that individuals do not have any pre-discursive consciousness and hence cannot be regarded as the origin of discursive changes (Howarth 2000: 44). This does not imply that the process of identity formation could be ignored. It rather requires us to reveal the forms of subjectivity that are produced by hegemonic discourses and that individuals take up or identify with through their articulations. For example, by stating that 'climate change is a myth' I could identify myself with the subjectivity of climate skeptics. Moreover, some subject positions are deeply institutionalized and hence cannot be that easily adopted by individual actors. For example, by stating that 'climate change is a security problem' I could not constitute myself as a security professional in the same way as I could constitute myself as a climate skeptic in the previous example. Depending on the context, people draw on quite different subject positions: as climate scientist at a conference, as a mother at a parents' evening, as a 'white European' at a passport control.

Decisive for the role of actors in discursive struggles is hence not the individual person but rather the subject position taken up when making an articulation. Moreover, as meaning is always unstable and temporary, so is the meaning of subject positions (Žižek 1990: 125). For example, what it means to be a scientist, a single father or a member of a subculture is contingent and dependent on discursive struggles. The same holds true for the authority and legitimacy attributed to any subject position. Scientists might be highly respected in one cultural context but in another there might be suspicion against the truth claims of science; even highly institutionalized forms of subjectivity might be discredited or challenged: for example a national corruption scandal might discredit bureaucrats or government officials in a country. Understood in this way, poststructuralist discourse theory does not speak to a concentration on the single text or articulation but calls for studying the context, i.e. the discourses and institutions that provide the subject positions taken up by individuals in the case under investigation.

Furthermore, due to the central position of the securitizing actor in the Copenhagen School approach, the audience takes only a passive and marginal role.[10] From a poststructuralist perspective, however, even the dichotomization of speaker and audience is problematic. I want to exemplify the divisions between a speech act script and a poststructuralist script of discourse through recourse to two different models of communication. The Copenhagen School securitization approach, on the one hand, mirrors the script of a transfer model of communication (Shannon and Weaver 1949). At first sight, it appears as if the Copenhagen School, by following Austin and Searle, would challenge a transfer model of communication: an utterance does more than simply transferring a message from a sender to a receiver – it constitutes an issue as a security problem. Yet, on closer inspection, the actant structure of the transfer model is maintained: you have the active speaker of security, and you have the passive audience. The audience understands the securitizing move on the basis of a conventionalized code shared with the speaker, triggering the desired effect, e.g. fear, acceptance of extreme measures. This is a model in which the audience reacts only passively – by either rejecting or accepting a prior securitizing move. That being said, there is an inherent contradiction in the call for a more active role for the audience as made by Balzacq (2005) and others: the audience is by definition passive – it is *listening and reacting to* something. Sticking to a dichotomic actant structure of speaker/audience thus necessarily reproduces the passivity of some actors in the securitization process.

In contemporary communication studies, the distinction between speakers and audiences in communication processes has been challenged (Sager 2004).[11] In many cases the 'sender' of a message can hardly be detected. In addition, it is wrong to assume that some communication partners act while others simply react or reconstruct meaning. Rather, in communication situations the *re*action of the apparent listener is just as important to the meaning of an utterance as the speech act itself. Communication should hence be understood as the collective construction of a common sense by different communicative instances. In this model, then, there is no speaker and no audience any more – all actants likewise participate in the construction of a common referent object (Brinker and Sager 1989: 129). Communication is the production of communality (*communio* = lat. community; *communicare* = lat. to do something jointly).

A *metaphor*, which helps in visualizing the basic assumption of the model, is the Brazilian *round village* in the Amazon region (Sager 2004: 139–140). In these villages, the single huts are circularly arranged around a common village square. This implies that all huts are oriented towards the center and everybody has access to it. In line with this metaphor, the basic principle of communication is not the transfer of messages or performance of actions but rather the common orientation of actants towards a jointly constituted sense. In the center of the village people display something, visible to all the other villagers. As an example one could imagine that individual villagers each put some logs on a common fire in the village center. Some might contribute more wood than others and some might rearrange the logs so that the fire receives more oxygen.[12] In short, all communicative partners contribute to a common referent, while the form of their contributions can be very diverse (Sager 2000: 72–73). And even more importantly, the resulting product, the fire, is qualitatively different from the single acts. No single actor, even if it was the village elder or local leader, could alone decide the outcome of the construction process.

The metaphor of the round-village is particularly illustrative for a theory of securitization beyond the well-established speaker/audience dichotomy. First, it can help develop a theory of securitization that does without the dichotomy of speaker and audience but still accounts for agency. Second, it helps replace the securitizing move and its illocutionary force as the analytical focus of securitization. Instead, it is the common orientation of different actants towards a common governmental problem – the village center in the above-mentioned metaphor – that brings about securitization. And third, the model stresses the necessarily material nature of any discursive process of securitization: the establishment of discursive communities is a process, which includes political architecture, media of communication, material tools and technologies.

Along the lines of the round-village metaphor, I want to reconceptualize securitization as the emergence of security *discourse coalitions* (Hajer 1995, 2006). Maarten Hajer (1995: 95) introduced this term to study processes of discursive struggle. A discourse coalition refers to a group of actors from different subject positions, which share a certain set of narratives and storylines. Any political discourse, according to Hajer, comprises generative metaphors, opposing storylines and narratives that make sense of a problem at stake. Climate change discourse, for example, is made up of different opposing storylines – some construct global warming as a myth or propaganda, others see it mainly as an environmental problem, and still others construct it in terms of economics or development. Political storylines comprise a diagnostic as well as a prognostic dimension (von Lucke *et al.* 2014: 7). The diagnostic dimension identifies certain issues as governmental problems and helps explain complex phenomena as a matter of fact. The prognostic dimension refers to the governmental imperatives (Dillon 2010) that follow from such political diagnoses. In concrete discursive struggles, actors from different subject positions draw upon these storylines to articulate their own arguments and interpretations. This gives weight to the respective narratives, especially if actors from subject positions attributed with a high moral authority take them up. And it makes them even more attractive to other actors as well. The model of discourse coalitions is summarized in Figure 3.1.

The discourse coalition model stresses the co-determination between discursive structures and collective agency (Gottweis 2003). Any single actor might have his or her reasons to, for example, draw on climate change and security storylines. Some might do so to raise attention to the issue, as with environmental NGOs for example; others might intend to extend their own authority, such as in the case of security professionals in

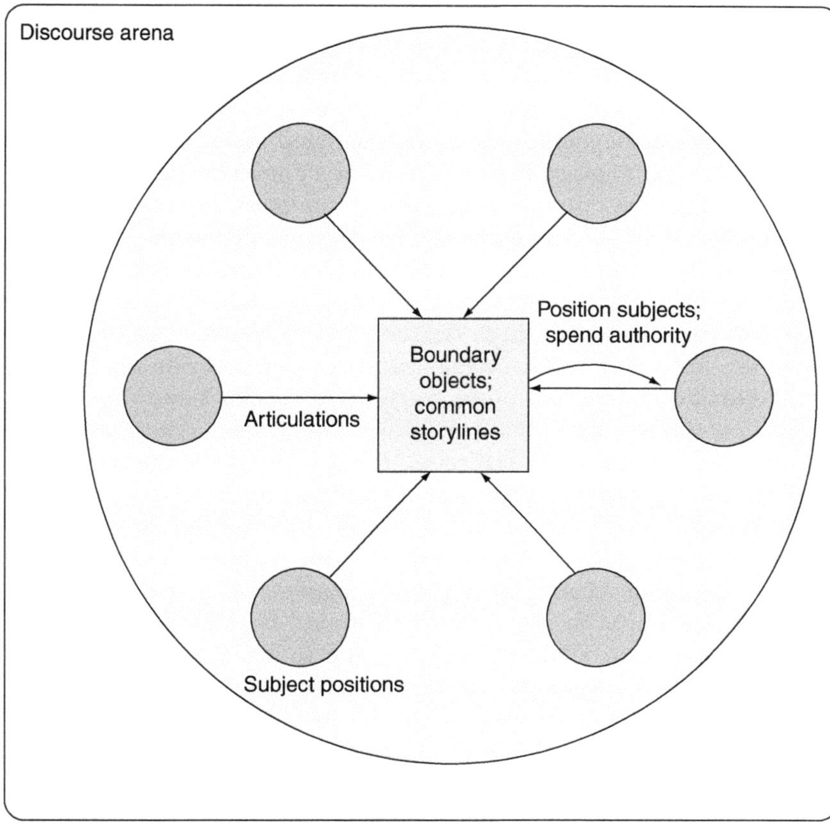

Figure 3.1 Discourse coalitions (source: author).

search of new fields of activity. Yet there are two qualifications that limit the importance of these individual strategies. First, the interests of actors depend on the subject positions that they adopt and these are provided by the discursive context (and its institutionalization). Second, what counts for the impact of any storyline is the cumulative level of discourse, and here individual intentions do not play a role anymore.[13] Discursive struggles, then, can be understood as competition between discourse coalitions over the right representation of an issue at stake (Hajer 1995: 59–60). Over time, through the process of a discursive struggle, there are two main possibilities for the success of one discourse coalition over others. Either more and more articulations are edged to the discursive outside, marginalized or discredited (as in the case of climate skeptics) so that one dominant reading remains, or one discourse coalition takes up more and more arguments, rationales and demands from other projects – with the same effect.

If one abandons the distinction between the securitizing actor and the audience in the Copenhagen School approach, however, one is forced to find a different criterion for a successful securitization than the acceptance by an audience. A solution is to leave behind the binary options of success and failure and to study the impact of securitization in gradual terms as discursive hegemony. Drawing on Maarten Hajer (1995: 60), I suggest two criteria to grasp the hegemony of a particular discourse or set of storylines. The first criterion is *discourse structuration*. Discourse structuration is highest if all agents of a discursive community have to draw upon the hegemonic storylines to make a relevant contribution to the discourse (Hajer 1995: 60). For example, the storyline of anthropogenic climate change today has reached a high degree of discourse structuration compared to the 1980s and 1990s. To make a reasonable contribution to international climate change discourse today, one cannot deny the human influence on the world's climate anymore. The second criterion for the influence of a certain securitization is *discourse institutionalization* (Hajer 1995: 61). Once accepted, security storylines become inscribed into societal and political practices, routines and organizations and materialize in important societal and governmental institutions. To give an example, the discourse on climate change as a *global management* problem, which was particularly prominent in the early 1990s, found its expression in global institutions such as the UNFCCC and attempts to install a global monitoring and management system for the flow of carbon.

Following the model developed in this section, a securitization of climate change, then, would require the evolution of a discourse coalition promoting storylines of climate change as a security issue. What holds a discourse coalition together is the common orientation towards a particular problem – such as dangerous climate change – as well as a set of storylines that make sense of this problem in particular ways. The success – or better, hegemony – of such a discourse coalition depends on the two criteria of discourse structuration, i.e. the degree to which the promoted storylines become sedimented and are taken for granted; and discourse institutionalization, i.e. the degree to which the storylines are inscribed into political institutions and programs.

Who's got the power?

Power is a crucial factor in any social science theory. However, it is seldom explicitly reflected in theoretical approaches. Critics have pointed to three different problems with regard to the notion of power in the Copenhagen School approach. Proponents of the Paris School have argued that securitization for the Copenhagen School is based on the "'social magic' power of language, a magic in which the conditions of possibility of threats are internal to the act of 'saying' security" (Balzacq 2010c: 1). Securitization, according to this critique, then takes the form of a self-referential practice:

securitizing moves constitute the realities they describe, thereby legitimizing a break with the ordinary (Buzan *et al.* 1998: Fn. 5). The power of language or discursive power in this respect means the ability of a speech act to induce changes in the discursive representation of reference objects. "In the concept of the Copenhagen School, any power of 'non-decision' [...] as well as more structural notions of power are not captured" (Stritzel 2007: 365).

In addition to the discursive power of language, Rita Taureck (2006: 17) has argued that Kenneth Waltz's notion of power as actor capabilities has influenced the thinking of the Copenhagen School.[14] In line with the neo-realist reasoning of Waltz, the social power of the securitizing actor is ascribed to its relational or systemic position (Buzan *et al.* 1998: 31).[15] Critics have argued that this has led to a concentration on securitizing actors from powerful positions (McDonald 2008: 574). Lastly, the security concept of the Copenhagen School, understood as the legitimization of exceptional measures, has been equated with sovereign power (Opitz 2008: 216).[16] While Opitz concludes that securitization might be a promising means to study how illiberal forms of government are inscribed into liberal regimes of government, others have criticized the equation of security with sovereign power. According to these critical voices, security discourses can take many different forms and security might even be rearticulated as human emancipation (Booth 1994).

From a poststructuralist take on power, drawing on the later Foucault and Laclau and Mouffe, these different critical remarks are very reasonable. A notion of discourse power and an understanding of power as relational capability are hardly compatible. In addition, a poststructuralist perspective would doubt the equating of security politics with exceptional measures and sovereign power. In the following I hence outline how processes of securitization are embedded within broader governmental landscapes, or topologies of power (see Collier 2009). These topologies of power are subject to a constant process of reconfiguration. This reconfigurational process is best understood through the notion of hegemony by Laclau and Mouffe (1985). Such a perspective on power can dissolve the above mentioned critique, as it both accounts for the productive power of discourse and also pays attention to how context (the broader governmental landscape) structures a discursive struggle – e.g. by providing subject positions that individuals might take up and by defining the horizon of possibility, the 'sayable' at any particular moment. Lastly, such an approach integrates sovereign power as one form of power amongst others into a broader governmental landscape and is hence open to the contingent political implications of securitization.

Foucault's productive power

With his concept of productive power, Foucault turned against a traditional understanding of sovereign power – in his eyes a totally deficient formal and purely juridical conception (Foucault 2005: 225) – or a notion of power as repression (Foucault 1978: 15–16).[17] A first characteristic of a Foucauldian understanding of power is that it is not regarded as a resource and hence cannot be owned by anybody. Rather, power is inscribed into all social relations. It is "a polyvalent force that runs through multiple sites in the social network" (Newman 2007: 51). And power is not related to an actor's intention (Digeser 1992: 983), as it seemingly is in the framework of the Copenhagen School. As power turns up in all social relations, it transcends society like a grid (Hindess 1996: 101). This points to a second main feature of a Foucauldian concept of power, namely its omnipresence: "Power is everywhere [...] because it comes from everywhere" (Foucault 1978: 93). This has to be understood less fatalistically as it seems at first sight, because power in Foucault's view is normatively neutral and does not necessarily work repressively. What this means exactly becomes clearer when we turn to a third feature of Foucault's concept of power: its productivity. Power does not only work in repressive ways, by prohibiting and excluding, but also as a productive force: "In fact power produces; it produces reality; it produces domains of objects and rituals of truth" (Foucault 1977: 194). The productivity of power on the one hand refers to the inseparable interrelation between power and knowledge (Foucault 1977: 27). Power enables the production of knowledge and at the same time the latter represents a source of power as it defines the horizon of possibilities in a given discursive field (Barnett and Duvall 2005: 55).[18] On the other hand, power is productive also with respect to the establishment of certain subjectivities, according to which subjects conduct – or govern – themselves. At the same time Foucault stresses that power can only be exercised on free subjects (Foucault 1982: 790). To be free according to Foucault means to have a *field of opportunities*. Other than in a classical account of power, freedom is not seen as the counterpart of power but as its prerequisite. To be a subject implies both being subject to something (the power relations that produce rationalities and subjectivities) and being a subject, i.e. a sovereign, self-acting person (Foucault 1982: 781). Consider the example of a green-minded citizen. On the one hand, this is a self-active agent, who decides to live in particular ways to safeguard the environment. On the other hand, he or she subjects her/himself to particular environmental discourses that define appropriate behavior. A neutral concept of power implies that power might foreclose choices and opportunities: to define the 'normal' or the 'sayable' also means to exclude the 'abnormal' by excluding, marginalizing or silencing certain perspectives on political issues. But it can also mean to produce choices and possibilities for a subject, e.g. by producing alternative discourses and behavioral norms, between which individuals might choose (see e.g. Hajer 1995).

These short remarks on Foucault's concept of power are able to qualify the notion of discursive power, or the power of language in the securitization approach. On the one hand, articulations are indeed related to power, insofar as they influence the thinkable and 'sayable,' and thereby construct a particular field of possibility. For example, articulations on climate change and security might bring about a discursive context in which military interventions in climate hotspot regions become thinkable or even appear reasonable. On the other hand, no actor possesses the power to speak security. What is more, the direction of power does not point unidirectionally from the speaker to the audience, as it is a polyvalent and multidirectional force. This points to a last analytical constriction of the Copenhagen School: the concentration on political elites, and macro-level discourses. The grand speeches of statesmen, the articulations of academic elites, and similar discourse events are at the center of attention for the Copenhagen School's framework of securitization. An approach inspired by Foucault would insist on *the multiple, everyday articulations* and discursive struggles, the mundane and often unnoticed practices of many different actors – or in the words of Jef Huysmans the "little security nothings" (Huysmans 2011) which together still can have a major impact.

The previous paragraphs have concentrated on Foucault's interest in the *microphysics* of power – its everyday occurrence and its pervasion of every small part of the social. At the same time, Foucault has also always been interested in the larger picture – in the interplay of politics, power and knowledge in political communities at the macro-level. His later thinking, starting with his seminal governmentality[19] lectures in 1977–1978 at the Collège de France (Foucault 2007, 2008), meant a twofold departure from his earlier approaches of political power. On the one hand, he began to focus on a particular, novel form of political power: government. For Foucault, government represents more than political steering. It refers to all practices that shape the conduct of people, from governing a household, governing a population, to the government of the self. Government as a form of shaping conduct has its roots in the "pastoral power" (Foucault 2007: 125–130) of the Christian religions, the secularization of which gave birth to a comprehensive economy of power in the seventeenth century. Government, furthermore, in this perspective, is always a reflexive practice as it presumes a certain problematization of the reality to be governed (or previous forms of government that are identified as inappropriate). Following Foucault in the genealogy of the modern state, the forms or 'arts' of government and related political discourses – including liberalism, biopolitics, or advanced liberal government – have been constantly changing (Foucault 2007: 247–248).

On the other hand, Foucault turned away from a totalizing analysis of all-encompassing power–knowledge formations. "Instead, increasingly he [...] examine[d] how existing elements are taken up and recombined" (Collier 2009: 90). This contradicts studies that apply a governmentality framework to diagnose a particular discourse, such as neoliberalism, as a totalizing force that shapes government mechanisms at all levels from the individual to the global in a coherent way. And it also speaks against works that diagnose a shift from one governmental logic to another – for example from liberal to neoliberal government. In this study, I draw on Collier's approach and understand power–knowledge relations as a complex topology of power:

> One technology of power may provide guiding norms and an orienting telos. But it does not saturate all power relations. Rather, it suggests a configurational principle that determines how heterogeneous elements – techniques, institutional arrangements, material forms and other technologies of power – are taken up and recombined.
>
> (Collier 2009: 89)

The concrete form and substance of government, in other words, is subject to a temporally and spatially specific (re-)combination of governmental rationalities and technologies. Such an approach has two advantages: first, it avoids the risk that analyses become mere self-fulfilling prophecies, if studied phenomena are simply subsumed under pre-given labels such as neoliberalism or biopolitics (see Rothe 2011). And, second, such a topological approach of power is in line with the discursive approach developed in the previous section, which argues that political meaning is contingent upon spatially and temporally specific discursive struggles (and the resulting establishment of discourse coalitions).

For a theory of securitization, such a topological analysis means that the concrete political form that a securitization process takes, and the particular forms that government here takes, is contingent upon the concrete discursive struggle in a given political space. Politics of security might take the form of sovereign power, but the latter is governmentalized, i.e. brought into a complex and changing relation with other forms of power and conducting people. This concrete form cannot be theoretically derived but only empirically determined in the concrete case.

Laclau and Mouffe's hegemony theory

Missing from Foucault's thinking on power, maybe because of his early death, is a concrete idea about how the microphysics of power relate to broader governmental structures (see Howarth 2000: 84). In terms of the topological approach put forward here, this refers to the question of how exactly different forms and rationales of government are articulated and combined to new forms of government. How exactly does the process of discursive struggle look that brings about shifts in the exercise of political power? And how can we make sense of the power at play in the process of recombination itself? A possibility to fill out this blind spot is provided by the discourse and hegemony theory of Ernesto Laclau and Chantal Mouffe.

Laclau and Mouffe draw on Foucault's notion of productive power and complement it with the concept of hegemony, which they take from a re-reading of the neo-Marxist thinker Antonio Gramsci (1971). Hegemony, in Gramsci's view, represents a form of power that combines the structural coercion of the relations of production with the construction of consent between social groups or classes. Hegemony is both coercion and consent (Gramsci 1971: 169). For a theory of securitization, hegemony is instructive, as the securitization of an issue requires both the construction of an existential threat and the construction of a valued referent object. And the latter necessarily requires the construction of consent in a political community – consent about the collective value and legitimacy of the referent object (Buzan *et al.* 1998: 25). Laclau and Mouffe (1985: 93–105) provide a means to study this construction of consent, not through single speech acts but through processes of discursive, or hegemonic, struggle.

Laclau and Mouffe linked Gramsci's thoughts on hegemony to a post-structuralist discourse theory. This meant concretely that they freed the concept of hegemony from its historical–materialist background and the related essentialism. They rejected the assumption that classes are given as universal political subjects and an antagonistic class struggle would be historically determined. Instead, they understood societal antagonisms as well as political subjects as rooted in contingent political discourse (Laclau and Mouffe 1985: 85–88). Political discourses can be regarded as relational system of very different societal needs and political demands. The smallest entity of a political discourse is hence the articulation of a political demand (Laclau 2005: 73). A hegemonic demand, then, is one that does not claim to meet a particular interest, but moreover claims to represent the common interest of a political community – or even the universal interest (Glynos and Howarth 2007: 116). Hence, the production of hegemony always implies a transformation of the particular and specific to the universal and general (Glynos and Howarth 2007: 116).

Recall what I said about the radical contingency of discourse in the previous section – discourses are never ultimately fixed. In political discourses, hegemony thus means two things: on the one hand, hegemony refers to those articulations that are consensually accepted as representing the common good (Nonhoff 2007: 180). On the other hand, those discursive elements are hegemonic that receive a taken-for-granted status, i.e. that are so widely shared that they appear as a naturally given (Hajer 1995: 61). A good example of a hegemonic discourse is neoclassical economics. Not only is the idea that there simply is no alternative to a liberal market economy commonly accepted, it is also a widely shared belief that the free interaction of market participants increases the public good, i.e. is a common or universal demand. Once they have become hegemonic, discourses hence follow a certain path-dependent development that will only fundamentally change in moments of dislocation.[20]

Laclau and Mouffe furthermore specify the discursive strategies[21] that provide the basis for a universalization of particular demands. The basis of any hegemonic project[22] is the construction of a chain of equivalence that unites different societal demands. This chain of equivalence is held together by one particular demand – the hegemonic demand or *empty signifier*[23] – that represents all the different demands (Nonhoff 2010: 42). A prerequisite of the construction of such a chain of equivalence is the existence of an antagonism – the binary division of the social space into two opposing camps (Nonhoff 2010: 43). On each side we can find the construction of a chain of equivalence – one representing the political demands, the other the threats, obstacles and grievances that must be overcome to realize these demands (Laclau and Mouffe 1985: 143–144). Political discourses are thus characterized by two different moments: the irreducible existence of social antagonisms and the promotion of the common good. Hegemony, from this perspective, stems from the production of consent between different social forces by a) constructing a common enemy (antagonism) and b) promising to enhance the common good of the inside group (Nonhoff 2010: 41–44). In reality every social and hence discursive space is traversed by a multiplicity of hegemonic struggles, in which different projects fight for discursive hegemony. In this diversity of struggles, quite logically, hegemonic demands can be promoted by different hegemonic projects. These signifiers, whose meaning is fiercely contested, are called *floating signifiers* – as they are floating between different projects and meanings (Laclau 2005: 131; Laclau and Mouffe 1985: 171).

Furthermore, one can analytically distinguish between an offensive and a defensive form of hegemony – which in empirical reality often overlap. According to Nonhoff, offensive hegemony refers to attempts by hegemonic projects to *challenge* and overcome *the existing hegemonic order* (Nonhoff 2007: 185). The structure that should be overcome is represented as an antagonistic enemy.[24] Hegemonic projects do so through a process of *coalition building*: "Hegemony can in the first place be seen as a political practice that involves the linking together of disparate demands to forge projects or 'discourse coalitions'" (Howarth 2010: 318). This coalition building requires "the construction of new discourses that can win over subjects to a particular project or coalition." The success of this discourse coalition or offensive hegemonic project depends on the acceptance of the promoted demands and storylines by broader parts of the political community. As Howarth argues, the influence of a project, and hence its chance to challenge the existing hegemony, especially requires that many socio-politically important subject positions join the coalition (Howarth 2010: 319). To give an *example*: consider a discourse coalition promoting the demand of climate security, understood as security from negative implications of climate change. An offensive hegemonic strategy would then imply presenting climate security as a universal demand, which not only represents everybody's interest but also substitutes other political demands, for example for the protection of the environment, for energy security, for public health, etc. The climate security demand could then be articulated to challenge, for example, the existing security discourse focusing on nation states and the military.

Defensive hegemony, on the contrary, is a form of rule that tries to secure the existence and survival of existing hegemonic structures (existing forms of knowledge, power relations and institutions). This form of power consists of attempts to channel grievances and societal demands by consensual narratives into forms that do not challenge existing orders in a fundamental way, e.g. through institutions of the welfare state (Nonhoff 2010: 185). While offensive hegemony draws on a logic of equivalence to construct two antagonistic camps, defensive hegemony is based on a logic of difference (Howarth 2010: 320–322). Existing grievances and resulting demands are kept separate, which prevents the establishment of discourse coalitions around collective demands in the first place. A similar possibility to prevent discourse coalitions from challenging the existing order in fundamental ways is to enter the respective discourse coalition and adopt some of the promoted demands. This means that the dominant governmental routines and practices need to be adopted to meet these new demands without changing in fundamental ways. At the same time this means that the discourse coalition and the related equivalential chain is broken up: some parts find their way into the existing hegemony while more radical demands, and the actors that promote them, are marginalized. A good example is how existing hegemonies have dealt with different ecological political projects: on the one hand the demand for environmental protection is taken up in existing forms of government, for example through the rationality of sustainability, while more radical calls for change, i.e. a radical transformation of a fossil fuels-based political economy, have been marginalized (see Swyngedouw 2010).

With these two forms of hegemony, we have found a promising toolkit with which to study the linkages between the microphysics of power (multiple discursive struggles) and broader (hegemonic) governmentalities. Through processes of coalition building, single discursive elements are fused into broader consensual storylines and rationales. The latter, if they reach a high degree of discourse structuration and institutionalization, become inscribed into the governmental landscape of a given policy field. On the other hand, hegemonic discourses and institutions that keep political demands and subjectivities distinct and prevent the establishment of discourse coalitions, such as the liberal institutions of the welfare state, can be regarded as strategies to sustain the existing hegemonic order (see Figure 3.2).

In this model, securitization represents a form of power in a twofold sense. On the one hand, securitization can take the form of the establishment of discourse coalitions, which bring together a variety of societal demands by articulating security as an empty signifier that unites them.[25]

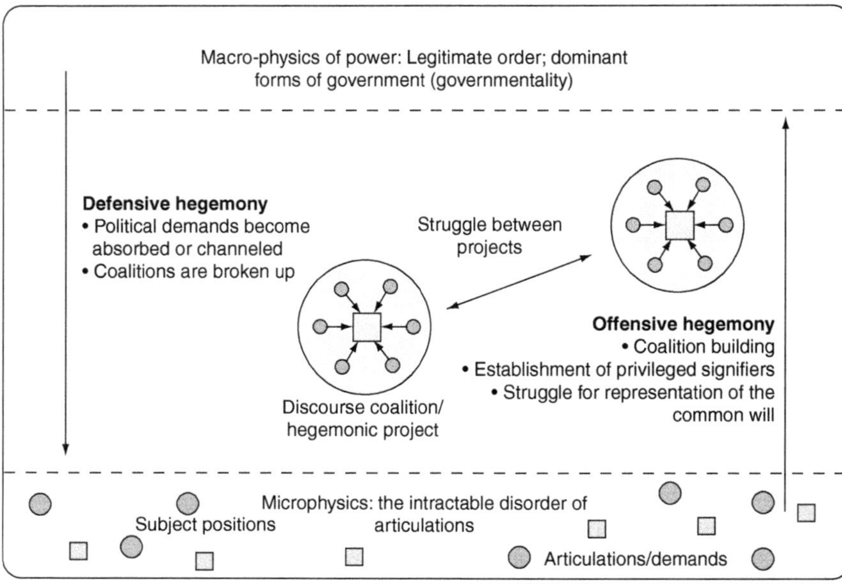

Figure 3.2 Hegemony and levels of political power (source: author).

Power here does not rest in the capabilities of securitizing actors, nor does it stem from the illocutionary force of security articulations. Rather, power emerges in the process of forging coalitions and by producing consent between different societal actors from opposing subject positions. What is more, any offensive hegemonic project draws on the construction of a political antagonism and securitization represents the discursive strategy behind such constructions. For example, radical ecological projects might construct a fossil fuels economy as an existential threat to humanity. Last, the articulation of security represents a powerful discursive weapon, which adds weight and a degree of urgency to the popular demands of the discourse coalition (see Rothe 2012).

On the other hand, securitization can have power effects by stabilizing an existing hegemony. This is the case if existing hegemonic discourses draw on security rationales to break up offensive hegemonic projects by keeping their demands distinct. Consider the case that security rationales are applied to distinguish between legitimate and illegitimate, priority and non-priority demands of ecological projects. This could mean that storylines on climate-induced migration or resource conflicts are taken up in governmental discourses, while more radical demands, e.g. for a fundamental transformation of the political economy, are excluded.

In sum, Laclau and Mouffe's hegemony theory provides a missing link between those identifying securitization processes with the micro-practices of security actors – by following Foucault's microphysics of productive power – and those studying it as a core political macro-discourse, able to create political antagonisms and suspend the normal political process. While such an approach remains "attuned to mess" (Squire 2013) and acknowledges the multiplicity of securitizing practices and struggles on the ground, it does not risk depoliticizing such mundane practices by tracing how they contribute to broader discursive, and hence hegemonic, struggles.

From Schmitt to Arendt – the political in securitization theory

In the Copenhagen School approach, securitization stands in an ambivalent relation to processes of *politicization*. Buzan *et al.* had assumed that one could distinguish between different degrees of urgency that characterize a political issue. Political problems are located on a continuum between 'apolitical' (there is no necessity to deal with the issue politically), politicized (the problems are being treated within the established rules of political decision making) to securitized (the problems are treated in emergency mode – breaking of rules is tolerated) (Buzan *et al.* 1998: 29). Once the issue crosses the threshold to the realm of security, it is moved out of normal politics and depoliticized (Buzan *et al.* 1998: 29).

Different scholars have critically addressed several points of this understanding of security and politics. On the one hand, the Schmittian legacy in the Copenhagen School approach has been criticized (Vuori 2008; Williams 2003). Critics have been especially skeptical of the framework's focus on the decisionist moment, at which the political sovereign through its securitizing move takes an issue out of politics. Second, the dichotomous picture of normal politics on the one hand and security on the other was questioned (Booth 2007: 165; Guzzini 2011: 332; McDonald 2008: 567). As Ole Wæver (2015: 122) recently argued, the Copenhagen School does not draw at all from a Schmittian concept of politics. Rather, it revolves around a Schmittian notion of security and an Arendtian concept of the political: "the place of security in the theory is as an anti-politics or the politically constituted limit to politics" (Wæver 2011: 478, endnote 472). In a nutshell, for Buzan *et al.* an issue is political as long as it remains in the public political sphere marked by open discourse on alternating perspectives. Security, then, marks the end of this debate: in security politics there is no alternative and the sovereign is able to adopt measures that would not be possible in 'normal' politics. Let me briefly introduce to you the political thinking of Carl Schmitt to better understand this distinction, which is crucial for any account of securitization.

Carl Schmitt and the political in securitization

For Carl Schmitt, ultra-conservative legal scholar and political philosopher with close affinities to the German Nazis in the 1930s – the existence of existential threats and the radical enmity between friends and foes is what constitutes the political. A political community can only fully constitute itself against the backdrop of a threat that existentially threatens the community (Williams 2003: 517). The possibility of emergency action and the breaking with the established normative/legal order for Schmitt directly follows from this constitutive antagonism (Schmitt 1996: 38–39). The threat that political antagonisms heat up and reach a degree of violent enmity cannot be suppressed by any political or legal order. The possibility of being overthrown is a constitutive feature of every order – which also applies to the legal order of a nation state (Huysmans 2006: 136). In this situation order can only be guaranteed by a political sovereign, who has the right and the duty to temporarily suspend the existing rules to get a hold of the situation. This can be called the decisionist moment in Schmitt's political thinking: it is the political sovereign who decides the appropriate moment for a suspension of existing rules (Schmitt 1993: 6–7). And, for Schmitt, the basis of the sovereign's decision is the identification of an external threat to the inner order and the related demarcation between friend and enemy. As the identity of the political community depends on such antagonisms, the sovereign's ability to decide the exception (which depends on the acceptance of society) and the friend–enemy distinction are mutually reinforcing (Williams 2003: 518).

As becomes apparent from this short outline of Schmitt's thought, the illocutionary event of the securitizing move – taking an issue out of politics into a 'security mode' – bears close resemblance to the decisionist moment in Schmitt's thinking. As I have outlined above, it was exactly this concentration on the moment of decision and the related dichotomization between 'the regular' and the exception, which has attracted much criticism (Williams 2003). This is supported by Schmitt's thinking itself. For Carl Schmitt, without the constitutive antagonism – the friend–enemy distinction – the political would cease to exist (Guzzini 2011: 332). Not only security or the state of exception but also 'regular' politics is marked by a friend–enemy distinction. According to him, the state of exception represents only the most radical form of political antagonism (Schmitt 1996: 38). Real politics, then, is marked by different degrees of intensity depending on how close it gets to this radical antagonism. "In other words, the political is based on the reality of difference and plurality and on the possibility of its politicization" (Odysseos and Petito 2009: 315). In a nutshell, what for the Copenhagen School characterizes security marks for Schmitt any political process – as argued by proponents of the approach themselves (Wæver 2011: 478).

What follows from the dichotomization of (Schmittian) security and (Arendtian) politics in the Copenhagen School approach is the necessity to provide clear criteria to distinguish between the two modes of social organization. Different scholars have pointed to the impossibility of such an endeavor (Huysmans 2006: 136). If one tries to define 'the breaking of rules' – according to the Copenhagen School, the defining feature of a security mode – on the basis of an existing order one finds oneself in a tautological trap. If the exception is what lets us identify the existing rules, we cannot draw on the existing rules to grasp the exception. Moreover, the definition is problematic as there is always the possibility of the exception becoming the rule.[26] In such situations, the breaking of rules as defining criterion for a securitization does not hold any more.

The only way out of this dilemma is to abandon the squishy concept of exceptional measures and with it the dichotomous differentiation between a 'regular' mode of politics and an exceptional security mode. To be able to grasp the political effects of securitization, I suggest drawing on Laclau and Mouffe's notion of the political. First, Laclau and Mouffe provide a reading of Schmitt's thinking, which accounts for the conflicting dimension of politics but avoids a dichotomization between politics and the exception. Second, it combines Schmitt's concept of politics with a cooperative understanding of politics in line with Hannah Arendt. According to such a productive concept of politics, it always includes the production of a common good – for example the security of a political community. And third, hegemony theory abandons the focus on the sovereign as a universal political subject – and in doing so allows the development of a theory of securitization in which political effects are not determined by the securitizing actor.

The political in discourse and hegemony theory

Schmitt's assumption that collective political identities require a common negative referent point such as through the construction of an enemy coincides with the poststructuralist assumption that every identity is based on difference. Hence a first characteristic of the political in discourse and hegemony theory is the existence of a constitutive antagonism – or the binary division of the social space. The relation between the inner political community and 'the other' is a paradoxical one, as the 'other' at the same time defines the community's identity and existentially threatens it (Nonhoff 2010: 39). Yet Laclau and Mouffe stress that the political does not only imply the existence of antagonism but also comprises striving for the common good. Within the political community, a fight over the right representation of the constitutive antagonism arises. Opposing political projects promote different arguments for the enhancement of the common good and define different enemies or obstacles and ways to overcome them (Mouffe 2005: 11–15).

A third feature of the political, thus, is the construction of a political community. Laclau demonstrates this with the case of populist discourses. For Laclau, *populism* is not a political abnormality but the logic of the political itself (Laclau 2005: 117). Populism according to him it is not defined by particular content but by a *specific ontological form* of political discourse (Žižek 2006). This ontological form refers to the construction of a chain of equivalence between different societal demands linked with the claim of one particular political demand (the empty signifier) to represent *the common* (will). On the basis of this claim to represent the common, a universal collective identity (the people) is constructed (Laclau 2005: 74). This implies dividing the social space into two camps: the inner community/the people on the one hand and a vague, external enemy – the corrupt elites, the foreigner, etc. – which is blamed for blocking the realization of the common will of the inside community (Laclau 2005: 93).

The competition between political projects, which each claim to represent the common will of the community, hence makes up the political. In line with this argumentation, a politicizing take on a specific issue is one that equals the logic of populism or offensive hegemony as outlined above. Politics requires the existence of clear political alternatives, which can enter into the political struggle (Mouffe 2005: 24–25). As these decisions are undecidable (i.e. there are no *objective* reasons which support one or the other option), they are intrinsically linked with power. This also holds true for consensual decisions, as consent can only be achieved by excluding some perspectives and demands. The establishment of a universal consent, in the interest of every party, hence remains an illusion (Mouffe 2005: 18). *For a theory of securitization* these assumptions imply that securitization processes can have a politicizing effect as long as they open up discursive contestation, for example if articulations open up debates about the subject and appropriate means of security politics.

Depoliticization, on the contrary, equated by Laclau and Mouffe to a logic of defensive hegemony (see previous section), can be understood as a process of the disappearance or suppression of political alternatives. It diminishes the possibility to choose or decide between different political and societal ideas. Depoliticization, hence, does not mean that an issue disappears from the governmental agenda or from the political process. Depoliticization refers to the way in which a particular issue is treated. Political management, technocratic administration or governance, and forms of government in which political contestation is substituted by seemingly consensual and technical forms of social regulation all represent a form of depoliticization. The reason is that such consensus-oriented forms of political steering undermine democratic decisions between distinct political alternatives (Mouffe 2005: 40–41). *For a theory of securitization* this implies that processes of securitization can have a depoliticizing effect if they foreclose discursive struggles and put political decisions in the hands of technocratic experts. This is for example the case if security risks and measures to deal with them are increasingly identified by a range of public or private risk-experts and hence taken out of public debate.

However, according to Slavoj Žižek (2006) there is another possibility of depoliticization that transcends the two logics outlined by Laclau and Mouffe. Žižek argued that a dichotomous differentiation between populist politicization on the one hand and post-political administration on the other is not adequate. Concretely, he criticized Laclau for having disregarded two characteristics of populism, which distinguishes it from other political projects. First, despite the ambiguity or vagueness that marks the constructed external enemy, in populist discourses it is always imagined as a *concrete figure* – for example a specific social group (the Jews, the bankers, the welfare scroungers, etc.). The social antagonism, which marks the field of the political in populist discourses, is reified through the figure of the *enemy* (Žižek 2006: 557). Second, for populists the failure made responsible for the perceived lack never lies in the system itself but in a part that does not function properly – a pathogenic symptom – which is constructed as an *external invader* that corrupts the system (Žižek 2006: 555). Depoliticized governance according to Žižek (2006: 567) is often legitimized through populist antagonisms, as these promote consensual/technocratic framings by blurring inner contradictions and constructing a common political will. The inner contradictions that characterize the inside community are obscured while the respective community is constructed as a community of fate (see Figure 3.3).

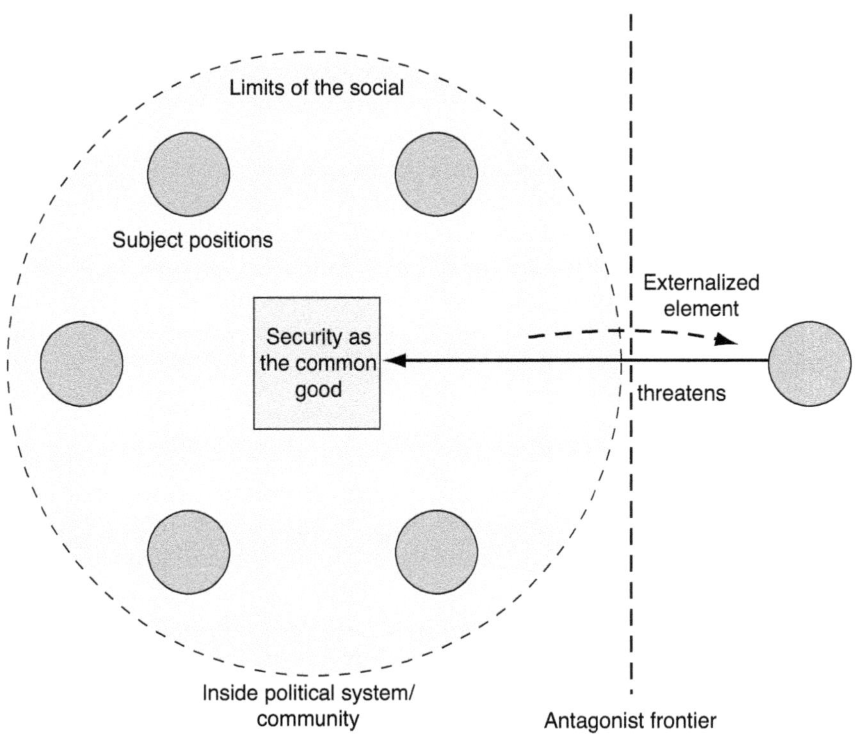

Figure 3.3 Securitization as post-political populism (source: author).

Such discourses, in which an existing hegemonic order is secured through recourse to populist narratives, can be called *post-political populism* (Swyngedouw 2010). A good example for such a post-political populism is the dominant discourse in the aftermath of the global economic crisis in 2008. The greedy and corrupt banker/manager was likewise identified as a personified enemy and as a pathogenic element that corrupted an otherwise functioning economic system.

Rather than is assumed by the Copenhagen School, I hence argue that the political effect of securitization is contingent and depends on the concrete form of the discursive struggle and the question of how political antagonisms become constructed in these struggles. Securitization processes can have a politicizing effect if they open up discursive contestation between identifiable political alternatives or bring in novel understandings that challenge the established security discourse. Attempts to reformulate or adopt concepts in a hegemonic security discourse, such as vulnerability, in the name of the people might have such a politicizing effect. A second face of securitization, however, is linked with the stabilization of the existing order and the silencing of political alternatives and hence equals a form of *depoliticization*. Hegemonic security discourses identify single demands, such as organized crime or the management of international migration, as common problems of the political community, which are then subjected to the governmental machinery of seemingly consensual administration. However, both depoliticization and politicization are gradual processes and security does not mark the endpoint of this trend – the political can never be fully suspended.

Last but not least, securitization can have a depoliticizing effect if it takes the form of *post-political populism*. This form of securitization is depoliticizing as it blurs the root causes of problems in social, political or economic structures, and it is depoliticizing as it silences political alternatives to a security framing of issues. Governance instruments seemingly follow functionally from the identified threat and appear as inevitable given the urgency of the issue.

Security, risk, resilience

In the last couple of years, more and more studies have proven that contemporary security policy increasingly focuses on assessing and managing a range of risks below the level of full-scale military threats. The Copenhagen School's framework of securitization, however, does not account for security conceptions below the threshold of existentialism. According to them, security has to be clearly distinguished from the field of internal security and its practices of risk management. Otherwise, one would risk diluting the conceptual coherence of security (Buzan *et al.* 1998: 21). Fixing the form of security as a conventional speech act was a particular reaction to the widening debate at the time on the development of the securitization approach: "Until the invention of the concept of securitization, 'widening security' had to specify either the actor (the state) or the sector (military), or else risk the 'everything becomes security' trap" (Wæver 2011: 469). Giving up every fixation of meaning, according to this argument, would pose serious methodological problems, as one would not be able to identify security anymore unless one would take either the actors or the sector of security as fixed.

This desire to keep security and risk distinct is perfectly expressed in an article by Ole Wæver, in which he acknowledges changes in security policy that give diffuse and anonymous risks and their management a much more prominent role compared to classical military threats. But, according to him, this does not lead to the conclusion that a theory of securitization had to include risk. Wæver (2011: 474) asks himself: "Did securitization theory maybe appropriately capture only the politics of security of 1995 or 1985?" He answers this rhetorical question by arguing that

> the question is not whether one practice changes into another (what would this mean?) Rather, the issue is that the utility and power of one theoretical model retracts while another expands. Securitization is securitization, and risk is risk [...]. This kind of change, therefore, does not challenge the consistency or precision of the theory – 'only' its relevance.
>
> (Wæver 2011: 474)

For its ignorance towards security practices below the threshold of exceptionalism, the Copenhagen School has earned much criticism (see e.g. Aradau *et al.* 2008: 149; Elbe 2008: 177–179; Salter 2008a: 244). A poststructuralist perspective supports this criticism. The problem is that, on the one hand, the Copenhagen School promotes a discursive ontology of security in line with a radically contingent concept of meaning (Hansen 2011: 359). Yet the radical contingency of meaning here is subject to one restriction: it is limited to the referent object of security, while the 'grammar of security' – the orientation towards existential threats and exceptional measures – remains fixed. A consequently poststructuralist perspective, on the contrary, has to stress the possibility of changes within the security discourse itself. This includes the establishment of new rationales and technologies – such as through the growing role of risk and resilience.

Risk as a discourse and a technology of government

Despite the Copenhagen School's ignorance of risk as a feature of security politics, the notion that international security politics is increasingly characterized by a focus on risk management today is quite commonplace (see e.g. Hameiri and Kühn 2011). Risk represents a particular technique for structuring the world, a practice that produces a specific knowledge about the future based on the calculation of probabilities and thereby enables certain political interventions (Rose 2001: 7).[27] Risk is a technology of knowledge production and a political practice to govern potential harms (to a population). Accordingly, there are no natural risks; everything can be a risk (Salter 2008a: 248). The concrete form of risk assessment and management, however, depends on the broader governmental discourses it is embedded within and hence governmental practices of risk management changed over time.

According to Foucault and his disciples such as Francois Ewald, risk technologies played a crucial role in the development of *modern government* (Ewald 1991). Modern government evolved when the use of political power turned away from the person of the sovereign or the territory as its main object or goal and instead focused on the population as an object that could be regulated and controlled (Foucault 1988: 161). The development of statistics and the so called probabilistic revolution in natural and human sciences made it possible to intervene into and govern the human life of people (Patomäki 2015: 129–130). Government turned away from the state's territory to the population with all its regularities and potentially dangerous deviances as the primary governmental object. Probabilistic risk became a governmental resource (Salter 2008a: 248). To describe this government of life-processes, Foucault used the term *biopolitics*. Dillon suggests thinking of biopolitics as the regulation of circulation (Dillon 2007: 3–4). The circulation (of people, goods, viruses, etc.) has to be monitored, controlled and managed to prevent harmful and promote 'good' circulation.

Biopolitics as the surveillance and regulation of circulation has been massively increased through the rise of *liberalism* as a political rationality. Individual freedom as the main objective of liberalism also required security against societal risks such as illness, poverty or criminality. Thus, *mechanisms of security* evolved in the institutions of the welfare state as a correlate to liberal freedom (Foucault 2008: 65–67). The evolving security dispositif turned out to be a specific form of risk management, which is characterized by defining an average value in a bandwidth of deviations as the desirable norm (Bigo 2008: 99). Government in this respect does not aim at the overall reduction of a societal harm but to secure an optimal distribution (e.g. a certain annual influenza death rate) against strong deviations. The comprehensive welfare system established in many liberal democracies in the mid-twentieth century was a system of collective risk management based on the principle of insurance (Ewald 1991: 204–205).

However, the rise of neoliberal discourses from the 1980s onward came with a problematization of a perceived excessive activity of the state, especially in the collective regulation of societal risks (Foucault 2008: 247). The result of this problematization, according to Dean, was that different forms of *advanced liberal government* became prominent (Dean 1999: 153). The major consequence of this political transformation was the individualization and privatization of societal risks. This is based on technologies that seek to govern people at a distance by making individuals responsible for their own health, safety etc. or by empowering stakeholders and at-risk groups to actively minimize their own vulnerability (Dean 1999: 166).

The dislocative moment of 9/11 and the inability of security analysts to anticipate or even imagine the terrorist attacks marked a fundamental dislocation in security discourses. 9/11 and other unforeseen events, such as Hurricane Katrina, that completely hit the US by surprise and soon became a symbol of non-linear environmental change, or the unexpected onset of the so called Arab Spring in 2011, further eroded the belief of security actors in the statistical calculability of risks. This acceptance that it is impossible to calculate all sorts of future risks had two major implications for security governance. On the one hand, it fuelled the emergence of practices of anticipation based on creative imagination and speculation rather than probability calculation (De Goede 2008; Diprose *et al.* 2008: 268–269). Such anticipatory practices for example include the development of discontinuous scenarios, so-called 'wild cards' – instead of the most probable scenario one tries to imagine all possible future trends to increase preparedness (De Goede 2008: 158–159). On the other hand, the assumption of living in an age of complexity and radical uncertainty (Kavalski 2008) paved the way for the concept of (social-ecological) resilience to enter security policy (Bourbeau 2013; Cavelty *et al.* 2015). In a nutshell, a resilience paradigm (which will be further discussed in Chapters 5 to 7) holds that radical uncertainty poses a limit to governmental planning and intervention. Instead it promotes the self-capacity of vulnerable systems and community to recover from harmful events or external shocks and to adapt to changing environmental conditions. Hence, rather than trying to protect citizens from all potential harms, governments should invest in the adaptive capacities of people and communities so that these become able to deal with future threats themselves.

Security and risk: two very different animals?

If one acknowledges the need to include risk discourses and practices in a theoretical framework on securitization, the question of how to conceptualize the relation between security and risk arises. Many scholars of a risk approach paint the picture of a 'traditional' security approach, which has slowly been replaced by a logic of risk management after the end of the Cold War. Corry, for example, argued that securitization was recently supplemented by a similar speech act or construction of danger, namely 'riskification' (Corry 2012). In a similar manner, von Lucke *et al.* distinguish between security threats, which are existential and urgent, and risk, which is a future potential of a harm, which might be managed or mitigated (von

Lucke *et al.* 2014: 6). Instead of an immediate threat, risk discourse constructs the vulnerability of a rather diffuse referent object and identifies certain collectives, communities or spaces as at-risk-groups (van Munster 2005: 6). Political measures do not aim for the defense against an enemy or threat any more but rather for the internal capabilities of the reference object, i.e. its adaptive capacity or resilience. Instead of legitimizing the temporary suspension of existing rules the speech act thus legitimizes the adoption of precautionary measures and risk reduction programs (Diprose *et al.* 2008: 271; von Lucke *et al.* 2014: 6). These routinized forms of risk management are presented as the opposite of crisis management in the exceptional mode of security (van Munster 2005: 6–8).

I want to argue, however, that such an observation of a shift from geopolitical security to risk or biopolitical security (Dillon 2007: 5) falls short of reality. There is a rather theoretical argument and several empirical arguments against this. Theoretically, "an understanding of risk practices as riskification may sound appealing but encounters the same problem of discursive fixation as securitization theory" (Petersen 2011: 710). Empirically, there are several examples proving that a clear-cut distinction between risk and security does not hold. For example, as Michael Dillon and Luis Lobo-Guerrero argue, the liberal way of war (e.g. humanitarian interventions) as expression of military, exceptional security has always been underpinned by a logic of risk informed by the biopolitical desire to optimize desired forms of life and contain undesired ones (Dillon and Lobo-Guerrero 2008). Furthermore, there are different cases in which a securitization process works through a terminology of risk, as for example in the case of securitizing HIV/AIDS: "The combined effect of the convergence of risk and security in the securitization of HIV/AIDS, then, is to begin integrating the older institutions of sovereign power within a biopolitical economy of power" (Elbe 2008: 193).

Finally, as Aradau and van Munster have shown, as early as the Cold War security strategy involved risk and other methods to govern an uncertain future, so the picture of a shift from security to risk after the end of the Cold War does not hold empirical validation (Aradau and van Munster 2011: 17–22). To assume the existence of two independent logics of securitization and riskification is problematic as it blurs how security and risk rationales actually become fused in complex processes of securitization; this is not a process of a unidirectional transfer from exceptional security to mundane risk – from geopolitics to biopolitics – but rather one of mutual imbrication of both discourses. In recent years, for example, an interplay between environmental, financial and security discourses meant that the principle of resilience became one of the dominant rationales in all of these discursive fields (Walker and Cooper 2011).

I want to go a step further to argue that security and risk are actually two sides of the same governmental coin. Drawing on discourse and hegemony, I would argue that practices of risk management can be read as an expression of a logic of difference. For example, the risk-based insurance mechanisms of the classical liberal welfare state order individual subjects in differential classifications and calculations of risk potentials: Individuals have become 'dividuals' (Deleuze 1992). In other words, societal demands are translated into differential risks and addressed separately. The geopolitical logic of security, by contrast, constitutes a logic of equivalence in that it demarcates its boundaries against an external enemy, and so underpins the 'likeness' of its individuals. Thus, in the same way as equivalence and difference are constitutive of every (hegemonic) discourse, risk and security are two sides of the same coin.

Even the most technocratic forms of risk management cannot do without constructing political antagonisms that define desired and pathogenic forms of life or circulation (travelers versus irregular migrants, consumption goods versus drugs, migratory birds versus invasive species). Decisive to the question of which policies are promoted through securitization is how this antagonism is constructed. If the security discourse constructs a fierce enmity between two different political communities, this can result in the adoption of extreme measures and a political state that equals a Schmittian state of exception. Such radical friend–enemy distinctions can be found in many recent security discourses, from Nazi Germany's 'total war' to Cold War macro-securitizations of two universal and utterly contradicting ideologies (see e.g. Buzan and Wæver 2009: 260). The very same discursive strategy, however, can promote practices of risk management. This is the case if securitizations shift "the frontier of antagonism to the periphery of the social" (Laclau and Mouffe 1985: 130). In this case, the inner antagonisms between political projects that mark a political community are edged to the outside (Laclau 1990: 223). This prevents the establishment of political projects, and society is constituted as a pure space of difference – a community inhabited by differential subject positions (see Figure 3.4).

The intimate relationship between security and risk is perfectly mirrored by the fact that the emergence of the Western welfare state coincided with the binary framing of world politics through the Cold War, which constituted the West as a homogeneous space with a common political identity. It was against the backdrop of this macro-securitization of the USSR that class conflicts within Western nations could be partially overcome and societal grievances – with the welfare state – could be managed through a comprehensive dispositif of risk management.

It is important to note, however, that in reality the boundary between these two political discourses is always a porous one. Even the Cold War – which is a paradigmatic example of a 'macro-securitization' in the sense of the Copenhagen School – was simultaneously made governable through practices of risk management (Aradau and van Munster 2011: 18). The dawning nuclear catastrophe, the existential threat that made up the core of this macro-securitization, was not an observable and present threat but

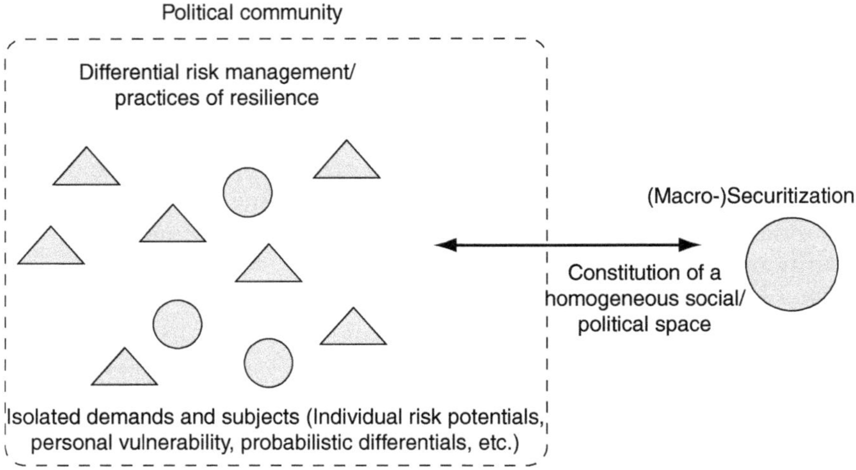

Figure 3.4 Risk and security as two sides of the same governmental coin (source: author).

a virtual entity in an uncertain future (Derrida *et al.* 1984). It was thus governed through practices and technologies such as scenario planning in order to make it thinkable and actionable (Aradau and van Munster 2011: 22–28) or measures to increase the preparedness of the people (Hagmann and Cavelty 2012: 83). Through dislocative moments such as the end of the Cold War or the terrorist attacks of 9/11, this already fragile boundary between security and risk logics became even more challenged (Aradau and Van Munster 2007: 89–90; De Goede 2008: 156–157).

These shifts did not mean that the logics of security and risk became completely equivalent. Rather, there is a complex interaction between the different rationalities and routines that mark each field. Instead of two distinct *dispositifs* of biopolitical and geopolitical security, there is an insecurity continuum that ranges from individual fears (such as fear of crime) to global apocalyptic discourses (Bigo 2001). Securitization, then, refers to discursive strategies that articulate these diverse fears, give them a concrete face and link them to particular forms of government. These cannot be a priori defined but depend on the concrete articulation, i.e. relationing, of risk and security in the respective case.

The metaphorical face of securitization

A last prominent point of the critique of the Copenhagen School refers to its narrow focus on language. Such critics have argued that the construction of security problems includes different media, such as pictures, visuals and artifacts (Hansen 2011), non-linguistic practices, as well as metaphors and other rhetorical tropes (Balzacq 2010c). To be fair, one has to acknowledge that the Copenhagen School also accounted for the possibility of securitization at a rather symbolic level. For example, Buzan *et al.* state in their 1998 volume that:

> it is important to note that the security speech act is not defined by uttering the word security [...] There will be instances in which the word security appears without this logic and other cases that operate according to that logic with only a metaphorical security reference.
>
> (Buzan *et al.* 1998: 27)

But this possibility of a securitization at the metaphorical level has not been followed up by the Copenhagen School – neither in its theoretical framework nor in the methodological categories that it provides. Approaches to account for the symbolic face of securitization remained scarce. Jef Huysmans for example accounts for metaphors in the securitization of migration – without attributing them a more prominent role (Huysmans 2000: 769). Lene Hansen has outlined a way to study visual securitization, i.e. the construction of security problems through images, pictures and symbols (Hansen 2011).[28] In this section I outline a possibility for integrating metaphors and emblematic symbols as important articulatory devices[29] into a theory securitization.

Most empirical studies drawing on the Copenhagen School's framework focused on the textual surface structure of language (the content of utterances – what has been said). According to discourse theory, below this surface structure, however, there is a second, linguistic deep structure. This discursive 'deep structure,' upon which speakers draw unconsciously in their statements, defines what can and cannot be said in certain discursive context (Lakoff and Johnson 1980). A crucial element of such a deep structure is symbols, icons, metaphors and similar rhetorical tropes. Metaphors and other tropes play a constitutive role in the construction of discourses by structuring and organizing knowledge (Howarth and Griggs 2008; Hülsse 2003; Walter and Helmig 2008). This is due to their feature of constructing analogies between referents to grasp complex phenomena through recourse to other, more accessible ones. Metaphorical expressions in language use are hence not only stylistic devices in speech and writing but actually the manifestations of deeper metaphorical structure of discourse (Hülsse and Spencer 2008: 578; Walter and Helmig 2008). Synonymy, metonymy, metaphors and other tropes are "not a pattern of thought, which give meaning to the social, but are part of the primal terrain, on which the social is constructed" (Sarasin 2003: 46, author's translation). Following Milliken, metaphor analysis is hence a promising means to study discourses and discursive changes (Milliken 1999: 235).

One possibility of translating the insights of metaphor theory for securitization is Jürgen Link's concept of the interdiscourse (Link 1984).

Drawing on Foucault's notion of discourse, Link assumes that societal mechanization and specialization led to the emergences of a range of specialized discourses. These are characterized by a specific discursive vocabulary, which is accessible only to a small number of experts. The discourse on climate change is a good example of such a specialized discourse. The total discursive space within a discursive community is made up of the totality of these specialized discourses as well as the field of the interdiscourse (Pêcheux 1984). The latter refers to the discursive field, which allows different specialized discourses to communicate with each other and with the general public on the basis of a shared discursive deep structure. According to Link, the distinctive feature of the interdiscourse is that its deep structure comprises a whole range of collective symbols (Link 1984: 67). These are commonplace signifiers that allow for a condensing of very complex issues into simple emblematic shortcuts and smaller storylines.

Collective symbols provide the interdiscourse with a figurative vocabulary to make sense of complex phenomena in an accessible way (Hülsse 2003: 222). Collective symbols allow one to construct analogies between multiple specialized discourses at the same time and hence construct affinities and coherence between them. For example, global climate change is constructed through metaphors of personification or the famous greenhouse metaphor. Such simplifying shortcuts represent the necessary condition for any interdiscursive exchange with other specialized areas such as development or security. Collective symbols, however, are not to be reduced to the phenomenon of the metaphor. Besides metaphors, they comprise tropes like metonymy, analogy, catachresis or synecdoche. But even certain material objects – as iconic signs – or pictures can function as collective symbols. In sum, collective symbols function as "boundary objects that help to translate and circulate knowledge. They synchronize research work, facilitate communication among different professional groups, and reinforce scientific authority" (Mayer 2012: 168). The prominent metaphor of the tipping point – the demarcation between a gradual and a chaotic climate change – as well as computer-based climate models in the climate change discourse are good examples for such boundary objects (Mayer 2012: 170–174). Collective symbols and boundary objects bring professionals from different subject positions together and help establish a shared problem perspective.

For a theory of securitization, this conception of discourse is important. First of all, it can be said that the interdiscourse represents the space in which a securitization takes place. On the one hand, securitization requires exchange between specialized discourses – for example, a security and a climate change discourse – each of which has its own discursive vocabulary, storylines and rules. On the other hand, I have argued above that a securitization requires a hegemonization of certain specialized demands by forging coalitions between different societal positions. The interdiscourse, then, provides the collective symbols and boundary objects that allow the communicating of specialized demands to a wider public. Collective symbols as boundary objects represent a powerful means to construct security storylines and to link different discursive domains. Put the other way around, the analysis of collective symbols and their relational structure is a good starting point to study securitization processes. If the same collective symbols are used in two different discourse domains, such as climate change and security, it becomes easier to link them argumentatively. For example, both environmental changes and the problem of migration are often framed through body/health metaphors. The resulting affinities between discursive domains, then, make it easier to argumentatively link the issues. The study of collective symbols over time is hence a powerful means to study discursive changes.

Finally, it is important to note that the concept of war and struggle itself is one of the most important collective symbols in Western interdiscourse. It takes the form of a *master metaphor*, which stabilizes and links different discursive fields. This means that metaphors of war and struggle are often applied to make specialized and complex phenomena such as global climate change comprehensible. The usage of collective symbols such as the war metaphor, however, has certain discursive effects, which might be called a metaphorical 'spill over.' If climate change is constructed along the lines of war metaphors, it is at the same time perceived in terms of war and struggle. This metaphorical spill over, I would call a *securitization precursor*. The usage of war metaphors in a discursive domain does not yet imply that the problem at stake is seen as a war, but it is constructed as an '*as-if war*.' The spillover hence produces certain discursive affinities which subsequently make it easier to link climate change with security arguments.

Ten theses on securitization

My aim in this chapter was to revise each of the theoretical building blocks, which could then make up the foundation of a consequent post-structuralist theory of securitization. Let me now try to sketch out the contours of such a revised concept of securitization. For this purpose I condense the findings made in this chapter into a few core assumptions or theses that together make up a comprehensive approach to securitization and already point to a possible operationalization of this framework.

Assumption 1: Securitization refers to processes of discursive struggle and not to a particular event or moment.

First, the concentration on speech acts as a means of securitization has been dismissed in favor of a comprehensive understanding of discourse, which stresses the articulatory and signifying character of all sorts of practices. Performativity is broadened to include both lingual and non-lingual practices of articulating security. In this sense, a computer model can articulate climate change as a security problem by relating it to other threats such as violent conflict. Securitization then does not take place at a particular moment; it is not a speech event, but requires ongoing practices of articulation in discursive struggles.

> *Assumption 2: A securitization of an issue requires the establishment of a discourse coalition that shares and rearticulates storylines of the respective problem as a security issue.*

In these discursive struggles, in which the right representation of political problems just as much as the appropriate means to cope with them are fought out, one cannot reasonably distinguish between speakers and audience. Instead the actant structure of securitization processes is conceptualized through the notion of discourse coalitions: heterogeneous groups of actors from different subject positions brought together by their common orientation towards a common problem perspective on a certain political issue.

> *Assumption 3: The political impact of a securitization project depends on the degree to which it establishes its demands and storylines as common sense (discourse structuration) and to which its claims become taken up by governmental institutions (discourse institutionalization).*

Abandoning the passive concept of the audience, however, required alternative criteria for the influence of securitization to be found beyond the binary possibilities of acceptance/non-acceptance by the audience. I thus argued for the evaluation of the influence of any securitization through two criteria: discourse structuration and discourse institutionalization. The first refers to the degree to which a discourse is commonly taken for granted. If the degree of structuration of a particular storyline is high, all discourse participants that want to make a reasonable contribution will have to draw on it. A good example is the storyline of anthropogenic climate change, which today is so commonly accepted that speakers denying it would be excluded from the dominant discourse. The second criterion is discourse institutionalization, referring to the degree to which discourses have become inscribed and materialized in social and political institutions.

With regard to the notion of power, a consequent theory of securitization holds that power is neither an intrinsic feature of (the performativity of) language nor is it an actor's capability. Instead power is an omnipresent relational force linked to the production of knowledge. So, power flows through discourses as these define the limits of the knowable. Power includes both dominance and the construction of consent.

Such a perspective helps us to identify the existing hegemonic security discourse at a given point of time and analyze how this discourse is linked with institutions, established practices and routines. I have introduced the term of governmental landscapes to account for the topological dimension of power – the fact that it becomes inscribed into material political arrangements or political technologies but that these arrangements and technologies are subject to constant reconfiguration and recombination.

> *Assumption 4: Securitization processes are embedded within broader governmental landscapes. Securitization then can be understood as a complex reconfiguration of these landscapes of power.*

For a theory of securitization, this implies that processes of securitization are embedded within broader governmental landscapes comprising the given established political technologies, rationalities and practices in the field under investigation. Securitization, then, refers to a process of reconfiguration of this governmental landscape. The introduction of security storylines into security discourse means that the established mechanisms, rationales and practices in the governmental landscape of climate politics become rearranged. For example, in the case of the UK, a securitization of climate change meant that adaptation policies were partly introduced into overall UK security policy based on an all-hazards approach.

> *Assumption 5: Securitization aims to produce consent and a common will of a political community. Thereby it either defends and stabilizes an existing hegemony or fundamentally challenges it, if applied by offensive hegemonic projects.*

Furthermore I argue that it becomes possible to study the reconfiguration of governmental landscapes on the basis of the two dimensions of *offensive as well as defensive hegemony*. Offensive hegemony, on the one hand, refers to the emergence of discourse coalitions that challenge the existing political discourses. Defensive hegemony, on the other hand, is to be found in those practices of the existing hegemony (the dominant discourses and governmental routines) that channel societal grievances and address single political demands so that the existing order remains untouched. Defensive hegemony is the reason for the inertia of existing governmental landscapes and can explain their tenacity. Securitization hence can take the form of an offensive hegemonic project in which the demand for security of a referent object (which has to be in the *common interest* – e.g. the national territory, the environment, the economy etc.) is constructed as a universal demand. This universal demand represents a whole series of further political demands: the security of the national economy could, for example, be related to political stability, social security, the maintenance of jobs, etc. *Securitization as offensive hegemony* aims to challenge the existing governmental rationalities and routines. For example, it is conceivable that climate change would be presented as a universal security threat in order to fundamentally challenge our fossil fuels-based economy and related policies. At the same time, however, there is a *defensive hegemonic variant of securitization*. This refers to the possibility that the security field is expanded in an almost unnoticed way. Governmental routines and rationales have expanded to new issues and single demands are captured and introduced to the existing machinery of security governance. Hegemonic discourses are not fixed entities; they absorb political contestation and take up single demands of offensive hegemonic discourse coalitions.

The concept of hegemony is closely linked to the concept of politics or the political. To overcome the binary distinction between an Arendtian ontology of politics and a realist or Schmittian ontology of security, my framework draws on Laclau and Mouffe's *notion of the political*. The construction of antagonisms here does not mark the end of regular politics but is rather one of its prerequisites. The competition between different political projects that each present their version of the social antagonism and a way for the realization of the common will makes up the heart of the political. The opposite of the political for Laclau and Mouffe is *the social*. The social refers to all those storylines, practices, routines, subject positions and rules that have become ingrained and are therefore not questioned any more. In short, the social comprises all those apparent facts (which in reality are sedimented discourses, e.g. the nation state, the institution of money, gender relations, etc.) that structure our everyday perception and behavior. As *depoliticization* we can thus understand all those practices that move an issue from the realm of the political to the social, by suppressing contestation, silencing alternative framings or by concealing the contingency of discourses so that these appear as a matter of fact. *Politicization*, on the contrary, means those practices that open up political contestation by challenging established interpretations and bringing in new political alternatives.

> *Assumption 6: Securitization does not necessarily mean an end to politics. It is to be understood in gradual trends of either politicization or depoliticization.*

Following this perspective I suggested understanding *securitization* not as a shift from politics to anti-politics but as a gradual trend of either politicization or depoliticization. Yes, securitizations *can have a depoliticizing effect*. Yet this is not restricted to a state of emergency in which the rules of the regular political process are suspended. Rather, I distinguished between different possible mechanisms of (de-)politicization.

One comes close to the Copenhagen School: security discourses can clearly have the potential to foreclose political struggle by silencing political alternatives. While many political demands are negotiable, the survival and integrity of the nation is not. Linking other demands to the security signifier can hence be an effective strategy of depoliticization.

Another depoliticizing mechanism of securitization was not discussed by the Copenhagen School: security discourse often comes along with modes of *technocratic governance*, leading to forms of *post-politics*. Security problems – such as climate change – are identified by academics and other experts, discussed by bureaucrats and military officials, and often addressed through technical means – which refers to technological instruments as well as expert driven forms of political management. This holds especially true for cooperative security problems, which are discussed and addressed in international expert fora, beyond any possibilities for democratic control or participation. Such technocratic and consensual forms of governance have an equally depoliticizing effect – they do not allow for any contestation or public dissent.

At the same time, *securitization can also have a politicizing effect*. This goes back to the fact that one necessary element of securitization – the construction of radical antagonism, the friend–enemy distinction – is at the heart of any political struggle. Security storylines can hence fuel offensive hegemonic political projects. Processes of securitization can bring the political back in. A right-wing populist discourse could for example politicize the discourse on migration by presenting a group of refugees as a threat to the security of their community. What this example further shows is that politicization is not always positive, whereas depoliticization is not necessarily negative.

Assumption 7: Securitization can take the form of post-political populism.

Lastly, *securitization can take the form of post-political populism*. Here, one part or group of the political system/community is externalized and presented as a threat which corrupts the system. Although it at first sight equals offensive hegemonic projects (it constructs an antagonism and a chain of equivalence), it actually has a depoliticizing effect. This is because it blurs the root causes of political problems, the fundamental cracks and inner contradictions of a system. Hence, post-political populism conceals the contradictions of an existing hegemonic discourse. Security discourses and fields can (but need not) play a crucial role in post-political populism. This is because the externalized, 'pathogenic' elements are often constructed along the vocabulary of the security discourse, and in extreme cases even subjected to repressive practices of security actors – think, for example, of the Jews in Nazi Germany.

> *Assumption 8: Risk and security are two sides of the same governmental coin. Processes of securitization themselves contribute to a growing amalgamation of the two.*

While the Copenhagen School mainly excluded the concept of risk and practices of risk management from a framework of securitization, I argue that it must be an intrinsic part of a comprehensive securitization theory. Many scholars have argued that security and risk are two different animals and follow different logics: security drawing on the identification of existential threats/enemies and the adoption of exceptional measures; risk revolving around mundane and routinized practices through which a whole range of risk factors are identified and mitigated. On the contrary, I argue that they are better to be understood as *two sides of the same governmental coin.*

> *Assumption 9: Securitizing discourse coalitions cannot only problematize novel issues as security relevant but can also bring about changes of the security discourse itself, e.g. by articulating alternative storylines on risk and resilience instead of military security.*

The boundary between these two sides, however, *has always been porous* and, in the aftermath of the end of the Cold War as well as 9/11, has become even more fragile. As a result the distinctions between inner and external security, between social control and security practices, became increasingly blurred. The ingrained and institutionalized security discourse became questioned, which opened up the space for alternative understandings of security, such as through the concept of risk. *The change of security discourses is intrinsically linked to processes of securitization* – the transfer of security discourses to other political fields such as the environment did not mean that these were unidirectionally subjected to a security logic. Rather, the existing rationales and concepts in these fields, among them concepts such as risk, vulnerability or resilience, also influenced the security discourse itself.

> *Assumption 10: Securitization takes place in the field of the interdiscourse, which provides a publicly shared set of collective symbols and boundary objects. Collective symbols become linked into broader narratives and storylines, which are used in discursive struggles to make sense of a particular political problem like climate change.*

Finally, a framework of securitization should be opened to means of securitization other than through the content structure of language. For example, the Copenhagen School already mentioned the possibility of a securitization taking place at a subtle, metaphorical level. Surprisingly, however metaphors and other symbolic means have seldom been included in empirical analyses of securitization processes. To fill out this blind spot, I introduced the notion of the *interdiscourse*. This is the discursive field, which allows different specialized discourses, such as the security or climate change discourses, to communicate with each other and to a broader audience. The basis is a discursive vocabulary, a common pool of *collective symbols*, which are shared by every member of the discourse community. Such collective symbols, which comprise figurative terms such as metaphors, analogies, metonymy and other rhetorical tropes as well as emblematic objects and pictures, allow one to link security with other specialized discourses such as the one on climate change. Collective symbols become linked in to broader meaningful narratives and storylines that are able to make sense of complex reality. Such storylines are the basis of political argumentation in discursive struggles about the right representation of political issues.

These ten assumptions, as the pillars of a revised poststructuralist theory of securitization, constitute the basis for my empirical investigation. They become further specified, elaborated and modified in the empirical investigation presented in Chapters 5 to 7.

Securitization meta-theoretically

One of the basic assumptions of the Copenhagen School was that the question of the meaning of security could only be empirically determined by studying the practices of actors that construct security problems. Surprisingly, as Balzacq acknowledges, the question of how to study these acts of securitization empirically has seldom been addressed (Balzacq 2010b: 31). Furthermore, while the Copenhagen School with the speech act approach provides a clear ontology of security, it does not give any idea about how these could be epistemologically grasped. Before we start searching for appropriate methods to study securitization, however understood, we have to develop an idea about which type of knowledge we can expect to produce with these methods.

The only way the Copenhagen School deals explicitly with philosophy of science is when it distinguishes itself from 'Critical Security Studies' (see Chapter 2) and other 'post-positivist' approaches in security studies. Such 'post-positivist' approaches, Buzan *et al.* (1998: 34) argued, focus mainly on the possibility (for the researcher) of changing established security discourses. The Copenhagen School instead argues for adopting a more realist perspective on security. According to Buzan *et al.*, it is solely the securitizing actor who decides what is a security issue. The analyst, then, is only able to study this process of securitization and to judge whether it is a good thing to securitize an issue or not:

> Although our philosophical position is in some sense more radically constructivist in holding security to always be a political construction

> and not something the analyst can describe as it 'really' is, in our purposes we are closer to traditional security studies, which at its best attempted to grasp security constellations and thereby steer them into benign interactions.
>
> (Buzan *et al.* 1998: 35)

In limiting meta-theoretical considerations to the question of whether it is possible for researchers to promote and change security concepts, the Copenhagen School does not substantially clarify their epistemological approach to securitization. Lately, the meta-theoretical debate on securitization theory has mainly been framed in terms of a schism between a constitutive and a causal model of securitization (Balzacq and Guzzini 2015: 99). While the former is associated with the speech act approach of the Copenhagen School, aiming to reveal how securitizing moves constitute certain realities, the Paris School formulates a more ambitious endeavor to causally explain securitizing processes by revealing the complex social mechanisms behind such processes (Balzacq 2010a: 66; Guzzini 2011: 329). I want to present a *tertium quid* as an alternative to these seemingly oppositional approaches, starting from the assumption that the distinction between constitution and causation, understanding and explanation does not hold up to closer inspection (Weber 1984: 28).

This third way requires abandoning a mechanistic understanding of causation, which dominates causal mechanism approaches to social science (Elster 1989: 3–4). Such a mechanistic notion of causality can also be found in the securitization framework of the Copenhagen School: "security talk is triggered by something, and once triggered by this something, it brings about certain de-democratizing or militarizing effects in a rather mechanical fashion" (Patomäki 2015: 129). The problem with a causal mechanism approach, and the Copenhagen School securitization framework, is that that they stick to a mechanistic notion (or 'billiard ball paradigm'[30]) of causality – it is possible to construct a causal sequence from the initial decision of an actor to argue for understanding an issue as a security problem, to the different means of articulation, and lastly the acceptance of the audience of the adopted exceptional measures. Such mechanistic notions of causality are deeply ingrained in our everyday thinking just as in political science:

> Causal explanations were from the start expressed in terms of impact, attraction, momentum – in short, pushes and pulls. This led to the notion that all causes were impacts, attractions, pushes and pulls, and all effects the result of pushes and pulls.
>
> (Hanson quoted from Glynos and Howarth 2007: 92)

I would argue that an approach of securitization has to be freed from the shackles of a mechanistic notion of causality. On the one hand, "the use of mechanisms is problematic in social science because it ignores how their meaning and significance is parasitic upon the practices in which they are embedded" (Glynos and Howarth 2007: 96). Translated into a theory of securitization, this implies that a focus on the mechanism of securitization – the securitizing speech act – ignores the social and political practices as well as broader discourses that make this particular articulation possible in the first place. On the other hand, such a mechanistic notion of causality draws on a clear-cut distinction between the explanans and the explanandum, which contradicts with the post-foundationalism of discourse theory.

As an alternative, I want to bring forward an understanding of emergent causality (Connolly 2005: 870). An understanding of *emergent causality* (Connolly 2004) distances itself from the idea of a linear cause–effect sequence and draws on the metaphors of resonance and spiral movement to make sense of causal relations. Understood as resonance, causality morphs into "energized complexities of mutual imbrication and interinvolvement, in which heretofore unconnected or loosely associated elements fold, bend, blend, emulsify, and dissolve into each other, forging a qualitative assemblage resistant to classical modes of explanation" (Connolly 2005: 870). While one cannot say that a particular security discourse X caused a particular policy Y, we can say that the interplay of different discourses and social practices can bring about and thereby cause something unforeseen and completely new. At the same time, the idea of spiral repetition stresses that changes or events are not necessarily caused by external forces, but causes may be immanent to the phenomenon at hand: "Sometimes that-which-repeats-itself also transforms itself. Because each iteration occurs in a unique context, each turn of the spiral enters into a new and distinctive assemblage" (Bennett 2001: 40). What is promoted here is an understanding of causality as gradual process, which is not defined by the simple presence, or absence, of factor X. Instead, causality is emerging, and can only be revealed after the fact.

The main interest of poststructuralist research lies in explaining the surprising, unexpected and wondrous by accounting for the *messiness and complexity* of the social world. Following Connolly, I argue that it is possible to *explain* the emergence of such wondrous phenomena while resisting the idea of full explicability and prediction (Connolly 2004: 343). To explain the process of a securitization of climate change and its policy implications, I suggest understanding securitization as a logic. A logic is an explanatory narrative that on the one hand allows for some generalization, i.e. is not limited to one particular context, and on the other hand respects the particularities of the case under investigation. "The logic of a discourse captures the rules that govern a meaningful practice, as well as the conditions that make the operation of such rules possible, whilst at the same time rendering them vulnerable to change" (Howarth 2010: 325). Understanding securitization as a logic, allows for the explaining of the tenacity of security discourses and studying processes of securitization as rule-following behavior. At the same time, the concept is open to radical projects that fundamentally challenge established notions of security. In contrast to laws, which are universally applicable and reproducible, a logic points to a "set of *family resemblances*" among different empirical phenomena (Glynos and Howarth 2007: 136).

A logic is a conceptual abstraction that is articulated differently in various contexts. Hence, by understanding securitization as a logic, one can avoid defining a particular script, mechanism or set of expected policy implications. According to this perspective, there is not a single logic of securitization that would apply to all cases studied. On the other hand, securitization is also not completely arbitrary or different in every case. Rather, securitization can follow different logics in multiple contexts, which nevertheless can be found in many different cases. To give an example, in some cases securitization processes could indeed follow a logic of exceptionality and lead to the adoption of extreme measures such as military violence. In other cases, however, securitization could follow a *logic of resilience* in which political problems are subjected to different forms of risk management (Corry 2014; Methmann and Oels 2015). In some cases the securitization process might follow a logic of discursive struggle, i.e. a public contestation about the subjects and meaning of security practices; in others securitization might follow a subtle logic of securitization through rather mundane and often unnoticed practices. All of these logics are instances of securitization processes as defined in the previous chapter, apply to a number of cases and yet are not completely generalizable, ahistorical or even universal. This also implies that different logics of securitization cannot be deduced from a theoretical framework but can only be identified through empirical analysis.

Using the framework in practice

To translate the assumptions of a poststructuralist theory of securitization into a methodical framework is the aim of this section. My overall method can be called a diachronic (or genealogical) discourse analysis. Simply put, discourse analysis means a detailed and structured analysis of texts (Balzacq 2010b: 41).[31] To be able to seize the whole complexity of security discourses, a range of empirical sources from media articles, through policy documents and speeches to qualitative interviews have been included in the analysis. These sources were studied over a longer period in order to be able to study changes over time. The analysis distinguished between an *early phase* of the securitization process from the early 1990s until the end of 2006 and a *peak phase* of securitization of climate change from 2007 to 2011. This distinction mainly served an analytical function. The analysis of the longer genealogy of climate security discourse was based on secondary literature as well as the results of conducted expert interviews. The main phase of the securitization process, on the contrary, was subjected to an in-depth discourse analysis.

A retroductive analysis (see Chapter 1) requires consideration of more than one type of empirical source (it thus bears resemblance to the idea of a data triangulation (see Denzin 1978: 302). The overall corpus of empirical material used in this analysis hence contains 107 newspaper articles from major international newspapers compiled through a systematic keyword database (lexis-nexis) search. Furthermore, 27 qualitative interviews were conducted with experts from different backgrounds. The selection of the interviewees followed a systematic theoretical sampling (Glaser and Strauss 1967: 45). On the basis of the explorative study I could identify the most important actor groups in the discourse. These were members of international governmental organizations (or national ministries in the UK case); of think tanks and NGOs; as well as of science.[32] In all cases three interviewees were chosen for each of these actor groups (3*3*3= 27), whereby I sought to include all relevant organizations in the field. The sample moreover comprised the totality of the speeches in the UNSC and UNGA debates on climate change and security (UNGA 2009; UNSC 2007a, b, 2011a, b). Last, I analyzed core policy documents and major think tank publications.[33] The number of sources included from each category was not determined a priori. Rather, I empirically identify the moment at which the findings could be regarded as *saturated* – namely, if the inclusion of further sources would not add new codes anymore but merely approve the already established categories and code clusters (Glaser and Strauss 1967: 63).

Practically speaking, this in-depth discourse analysis was based on a *semi-open coding* strategy (Glaser and Strauss 1967: 102–115), which was implemented practically with the help of a qualitative data analysis software (MAXQDA). The coding strategy was semi-open, as the basic codes were predetermined through the heuristic framework: these were storylines, collective symbols, political practices and measures, as well as subject positions. The different empirically enriched codes (such as war metaphors, the storyline of research conflicts, the subject positions of climate celebrities, etc.) were only inductively gathered from the empirical sources. At

Table 3.1 Different empirical sources and their function

All political levels	*International arena*	*Mediterranean*	*United Kingdom*
27 Interviews, 107 Newspaper articles	Main source: speeches at the UNSC and UNGA Additional sources: think tank publications and policy documents	Main sources: EU policy documents; think tank and NGO publications Additional: newspaper articles and interviews	Main sources: Core UK policy documents; interviews Additional: think tank and NGO publications

the same time, the analytical perspective was always open and aware of new concepts (through so-called in-vivo coding). Through the application of sub-codes and the construction of code families, it was possible to reconstruct the discursive structure in the empirical material in an open and unbiased way (Keller 2006: 105–106). According to a poststructuralist discourse theory, the discursive structure does not show up in the single text, but emerges as a structure between the analyzed texts and hence shows up as repetitive patterns. Hence the aim of empirical analysis is not to reconstruct the structures of meaning in single texts or sources. Discourse analysis rather assumes, first of all, that a single source can *never* cover the whole discursive field; second, that different discourses or discursive fields can show up in a single source; and third, that a single event or utterance (a speech act) cannot decide the discursive meaning on its own (Keller 2008: 275).

The empirical analysis applied in this study was hence not interested in the intratextual characteristics of single sources (e.g. the storyline of a text or its performative function). It focused on an *intertextual analysis* to reveal structures of meaning that show up in the material as repetitive patterns see (Balzacq 2010b: 43). Discourse analysis hence always requires a broader corpus of analyzed material. The codes that were identified in a first step of analysis, through an explorative study of a smaller corpus of materials, were re-applied in further steps of the retroductive circle, enabling a constant process of comparison (Glaser and Strauss 1967: 102–115). Thereby, possible nodal points, prominent subject positions, dominant storylines or prominent policy measures could be identified. This general methodical approach, which tried to translate the metatheoretical assumptions outlined above for empirical analysis, metaphorically speaking represented *the key ring* that held together the various methods or *keys to the empirical world* in a coherent way (Kruse *et al.* 2011: 62). The combination of different methods raised the plausibility of the developed explanations and also accounted for the complexity of the theoretical framework of securitization developed in this book.

The different levels of analysis derived from a poststructuralist theory of securitization as well as the units they comprise and adopted methods to grasp them are summarized in Figure 3.5. One can distinguish between three different analytical layers, each of which is approached through a particular analytical perspective. The first of these layers refers to the deep structure of the climate change and security discourse (or the interdiscourse between the specialized climate change and security discourses). This comprises the basic collective symbols, nodal points and prominent signifiers (i.e. categories and concepts) that mark the discourse at a particular point of time. This deep structure was studied through a metaphor-based discourse analysis. A second analytical layer refers to the different storylines and narratives. A narrative discourse analysis helps to reveal how the single discursive elements (metaphors, demands, etc.) are condensed

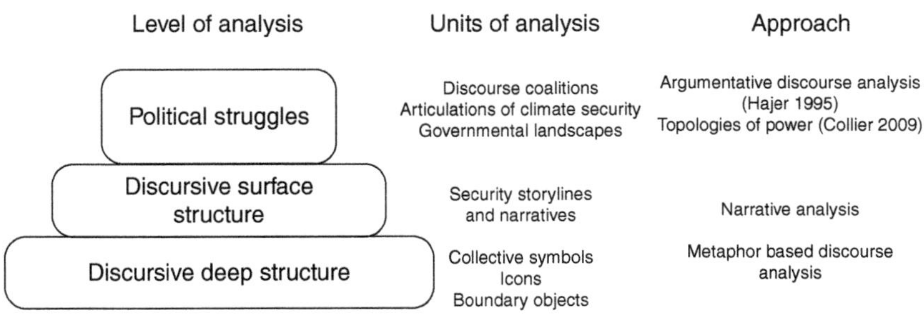

Figure 3.5 Levels, units and approaches in empirical analysis (source: author).

into broader political storylines. A final analytical layer is the level of concrete discursive struggles between competing political projects. I used Maarten Hajer's argumentative discourse analysis here to study how actors from different subject positions take up discursive storylines in processes of discursive struggle and how actors are forged into broader discourse coalitions or political projects.

Notes

1 Antinomy refers to the combination of assumptions in a theory that are absolutely logical taken by themselves, yet, taken together, construct a contradictory picture.

2 In a recent forum contribution Ole Wæver has defended his approach of drawing on diverse, partially opposing theories in developing the securitization model: "we all know numerous examples of internally consistent theories that draw on several traditions – and many more examples of theories that stay within their 'box' and yet are horribly inconsistent" (Wæver 2015: 124). While this remark is to some degree right, it is equally true that scholars have not only criticized the Copenhagen School for combining theories of different traditions but have also highlighted the contradictions that arise from this combination. Furthermore, despite all sympathy for theoretical pluralism, there are certain limits for the combination of theories. These limits are posed by the meta-theoretical, i.e. ontological as well as epistemological foundations of each theory. Given these limits it is actually problematic for the consistency of a framework to combine realism with poststructuralism, or a causal with a constitutive logic of inquiry.

3 The aim of this book is similar to Hansen's who argued for an extension of the Copenhagen School's framework to develop a more explicitly poststructuralist securitization theory (Hansen 2011). While the aim of Hansen is very similar to this book, the adopted approach is very different: Hansen draws on Foucault to suggest minor revisions and extensions of the framework, while this book seeks to conduct a thorough deconstruction of the framework and to develop a comprehensive theory of securitization.

4 Following de Saussure, the signs of a language system are related to each other along two different dimensions. Syntagmatic relations, on the one hand, refer to the possibility of signs being combined to form broader meaningful unities

like sentences (De Saussure 1967: 123–124). In the sentence 'Climate change is one of the major international security problems today' every sign or word determines the previous and following words. Associative or paradigmatic relations, on the other hand, unite signs of a particular class that in a sentence could stand as a substitute for each other. In the example sentence just given, 'climate change' could for instance be substituted with the signifiers 'terrorism,' 'piracy' or 'organized crime.'

5 A chair, for example, receives its meaning in a relational structure by being distinguished from other material objects while it is at the same time equated to others, for example with other seating furniture or other wooden objects.

6 In a religious discourse it might be equated with other divine punishments such as a bad harvest and differentiated from e.g. earthly phenomena, or from 'normal wind,' etc.

7 Laclau and Mouffe explain their argument with an example in which a builder and his apprentice are building a wall. The builder shouts to his apprentice 'block,' 'pillar,' 'slab' or 'beam' so he can pass him different types of building stones needed to construct the wall. Laclau and Mouffe argue that not only are the linguistic actions of shouting out 'block,' 'pillar,' etc. discursive elements but that non-linguistic elements such as passing the construction material and combining it in the right fashion are part of the discursive totality of building a wall (Laclau and Mouffe 1985: 108).

8 This distinction between philosophical and theoretical ontology goes back to Jackson (2011). The first refers to "the conceptual and philosophical basis on which claims about the world are formulated in the first place." Theoretical ontology, on the contrary, refers to "a catalog of objects, processes, and factors that a given line of research expects to exist" (Jackson 2011: 28).

9 In fact many empirical studies on securitization draw on methodological approaches of discourse analysis that are much more in line with the philosophical ontology of the Copenhagen School than with its concrete ontological concepts.

10 This even went so far as to argue that the audience could be completely disregarded (Floyd 2011).

11 It should be clear that communication is not to be confused with discourse. Discourse theory is precisely not interested in a goal-oriented use of language (such as the transfer of a message). A constitution model of communication, however, is compatible with discourse theory. And it provides a good method of analytically grasping processes of a discursive struggle about meaning.

12 This metaphor should be understood as an ideal type. In reality power relations change the setting, for example by silencing or excluding some positions.

13 Foucault calls this the "tactical polyvalency" of discourse (Foucault 1978: 101). Discourse coalitions do not follow one particular strategy. Rather, their influence goes back to the complex and non-determined cumulative result of individual strategies.

14 In contrast to Waltz, however, the Copenhagen School understands actor capabilities not solely in material terms. To account for non-material capabilities, the Copenhagen School draws on Bourdieu's broader notion of capital to argue that the success of a securitizing move depends on the social and cultural capital of the speaker (Buzan *et al.* 1998: Fn. 5).

15 Such a personal understanding of power as capability (Barnett and Duvall 2005: 45) is already present in the speech act theory. The power, or perlucatory effect, of a speech act rests upon a social convention. The same social convention defines certain social roles as the appropriate speakers of the speech acts. Only a registrar can perform a marriage, only a property owner can make a claim for his lost property, only sane adults can make a statement, etc.

16 In line with Foucault, I understand sovereign power as the power of the political sovereign that is exercised over a territory through the means of law and its enforcement through repressive state apparatuses (Foucault 1977: 7; 1978: 135).
17 Such a widespread notion of power understands the latter as the capability to enforce one's own interests against the will of others (Lukes 1974: 16–19). This notion of power is perfectly expressed in the following definition by Robert Dahl: "A has power over B to the extent that he can get B to do something that B would not otherwise do" (Dahl 1957: 202–203).
18 Technologies of power are the means through which power is exercised, i.e. the "apparatuses, practices, institutions, legal forms, etc. that allow the governing of subjects according to a political rationality" (Lemke 1997: 147, author's translation).
19 Foucault introduced the composite term governmentality (gouverner = to govern; mentalité = mentality) to grasp the power–knowledge nexus in modern government. In a broader sense, governmentality refers to the ways through which good government is reflected upon, i.e. all articulations at a particular historical moment that ask how certain problems are to be governed (Dean 1999: 16). In a more narrow understanding, governmentality refers to the process through which government, a form of power based on the conduct of people, evolved as the main modus of political rule (Dean 1999: 16.). In this study I will follow this narrow understanding of governmentality as a particular form of rule.
20 The term 'dislocation' refers to unforeseen events, such as a global economic crisis, which challenge the established discourse in a fundamental way (Laclau 1990).
21 Strategies here have to be understood as the non-determined agglomerate of individual intentions and tactics (Nonhoff 2007: 184).
22 One can speak of a hegemonic project if a coalition of actors, a discourse coalition in my terminology, links a universal demand with a claim of political supremacy (Nonhoff 2007).
23 The role of the hegemonic demand is filled out by a so-called *empty signifier*. As the signifier that represents the universal demand at the same time represents all the other demands out of the equivalential chain, it is filled with more and more meaning. This also means that the signifier becomes conceptually emptied: the more demand it represents, the more unclear and vague it becomes (Laclau 2005: 96).
24 Laclau hence associates populist discourse with a logic of equivalence (Laclau 2005: 78). Populist projects construct an antagonism between the people on the one hand and the representatives of the present system on the other. These political elites are then presented as an obstacle for the realization of the popular demands, i.e. the will of the people.
25 As Laclau argues the security signifier, similar to comparable signifiers such as freedom or justice, are especially suitable for functioning as empty signifiers in a populist strategy (Laclau 2005: 96). This is because, in contrast to other signifiers, they are always already empty as they do not have any positive content at all but give name to a perfection that is constitutively abstinent and unattainable. In other words, in a world without threats the concept of security would not exist.
26 Different authors have pointed to the fact that in the recent efforts to combat international terrorism by the US and the EU, control and surveillance measures that initially violated existing laws have already become the rule (Salter 2008b). Furthermore, one must not forget in this respect that a large share of nation states is ruled on the basis of emergency law (Neocleous 2006: 204).

27 In classical risk assessment, risk is a function of the probability of a future event combined with its negative impact/magnitude.
28 Other scholars have applied metaphor-analytical methods to study how security threats such as terrorism are socially constructed – however, without explicitly linking up to a securitization perspective (Hülsse and Spencer 2008).
29 As metaphors and other tropes such as analogies or synecdoche construct relations of equivalence between two referents, they are important means of articulation, i.e. practices of relating discursive elements along the axes of difference and equivalence (Hülsse 2003: 129).
30 Even if causalities in social reality are complex, according to this paradigm, in principle it is possible to trace back a series of events to their primary cause – just as in the case of billiards, in which different balls subsequently set each other in motion caused by an initial shot.
31 The present analysis, therefore, studies practices and other non-linguistic phenomena (such as social and political institutions) on the basis of text. While this might initially sound problematic, it is justified as there is actually no other way: as it was argued in the theory part of this book, the practices and material objects of the world that we perceive are always already mediated through language (Nonhoff 2011: 96). If you consider a researcher studying security practices through participatory observation, for example, she will formulate her results on the basis of a particular text – namely the notes she has taken down through the observation (actually her thoughts during the observation being already based on language). And this text is just as dependent on the discourses that surround the analyzed practices as, for example, secondary literature. This does not mean that her text is not of a different quality than secondary literature – it can be more detailed or might reveal new insights – but it is still a text. We could call this, in line with Denninger *et al.*, a "methodological primacy of discourse" (Denninger *et al.* 2010: 211).
32 A list of interviewees can be provided via email upon request.
33 These core documents and major publications were identified on the basis of the interviews and the explorative study of newspaper coverage.

References

Aradau, Claudia and Rens Van Munster. 2007. "Governing terrorism through risk: taking precautions,(un) knowing the future." *European Journal of International Relations* 13 (1): 89–115.

Aradau, Claudia and Rens van Munster. 2011. *Politics of catastrophe: genealogies of the unknown, PRIO New Security Studies.* London: Routledge.

Aradau, Claudia, Luis Lobo-Guerrero and Rens Van Munster. 2008. "Security, technologies of risk, and the political: guest editors' introduction." *Security Dialogue* 39 (2–3): 147–154.

Balzacq, Thierry. 2005. "The three faces of securitization: political agency, audience and context." *European Journal of International Relations* 11 (2): 171–201.

Balzacq, Thierry. 2010a. "Constructivism and securitization studies" In *The Routledge Handbook of Security Studies,* edited by Miriam Dunn Cavelty and Victor Maurer, 56–72. New York: Routledge.

Balzacq, Thierry. 2010b. "Enquiries into methods: a new framework for securitization analysis." In *Securitization theory: how security problems emerge and dissolve,* edited by Thierry Balzacq, 31–54. London, New York: Routledge.

Balzacq, Thierry. 2010c. "A theory of securitization: origins, core assumptions, and

variants." In *Securitization theory: how security problems emerge and dissolve,* edited by Thierry Balzacq, 1–30. London, New York: Routledge.
Balzacq, Thierry and Stefano Guzzini. 2015. "Introduction: 'What kind of theory–if any–is securitization?'." *International Relations* 29 (1): 97–102.
Barnett, Michael and Raymond Duvall. 2005. "Power in international politics." *International Organization* 59 (01): 39–75.
Bennett, Jane. 2001. *The enchantment of modern life: attachments, crossings, and ethics.* Princeton, NJ: Princeton University Press.
Bigo, Didier. 2001. "The Möbius ribbon of internal and external security(ies)." In *Identities, borders, orders: rethinking international relations theory,* edited by Mathias Albert, David Jacobson and Yosef Lapid, 91–116. Minneapolis, MN: University of Minnesota Press.
Bigo, Didier. 2008. "Security: a field left fallow." In *Foucault on politics, security and war,* edited by Andrew W. Neal and Michael Dillon, 93–114. London: Palgrave.
Booth, Ken. 1994. "Security and self reflections of a fallen realist." YCISS Occasional Paper 26, Aberystwyth.
Booth, Ken. 2007. *Theory of world security, Cambridge studies in international relations, 105.* Cambridge: CUP.
Bourbeau, Philippe. 2013. "Resiliencism: premises and promises in securitisation research." *Resilience* 1 (1): 3–17.
Brinker, Klaus and Sven Frederik Sager. 1989. *Linguistische Gesprächsanalyse. Eine Einführung, Grundlagen der Germanistik, 30.* Berlin: E. Schmidt.
Butler, Judith. 1993. *Bodies that matter: on the discursive limits of "sex."* 3rd edition. London: Routledge Chapman & Hall.
Butler, Judith. 1997. *Excitable speech: a politics of the performative.* 3rd edition. London: Routledge Chapman & Hall.
Buzan, Barry and Ole Wæver 2009. "Macrosecuritisation and security constellations: reconsidering scale in securitisation theory." *Review of International Studies* 35 (2): 253–276.
Buzan, Barry, Ole Wæver and Jap De Wilde. 1998. *Security: a new framework for analysis.* Boulder, CO: Lynne Rienner Publishers.
Cavelty, Myriam Dunn, Mareile Kaufmann and Kristian Søby Kristensen. 2015. "Resilience and (in)security: practices, subjects, temporalities." *Security Dialogue* 46 (1): 3–14.
Collier, Stephen J. 2009. "Topologies of power. Foucault's analysis of political government beyond 'governmentality'." *Theory, Culture and Society* 26 (6): 78–108.
Connolly, William E. 2004. "Method, problem, faith." In *Problems and methods in the study of politics,* edited by Ian Shapiro, Rogers M. Smith and Tarek E. Masoud, 332–349. Cambridge: CUP.
Connolly, William E. 2005. "The evangelical–capitalist resonance machine." *Political Theory* 33 (6): 869–886.
Corry, Olaf. 2012. "Securitisation and 'riskification': second-order security and the politics of climate change." *Millennium-Journal of International Studies* 40 (2): 235–258.
Corry, Olaf. 2014. "From defense to resilience: environmental security beyond neoliberalism." *International Political Sociology* 8 (3): 256–274.
Dahl, Robert A. 1957. "The concept of power." *Behavioral science* 2 (3): 201–215.
De Goede, Marieke. 2008. "Beyond risk: premediation and the post-9/11 security imagination." *Security Dialogue* 39 (2–3): 155–176.

De Saussure, Ferdinand. 1967. *Grundfragen der allgemeinen Sprachwissenschaft.* Berlin: Walter de Gruyter.

Dean, Mitchell. 1999. *Governmentality.* London: Sage Publications.

Deleuze, Gilles. 1992. "Postscript on the societies of control." *October* 59 (4): 3–7.

Denninger, Tina, Silke van Dyk, Stephan Lessenich and Anna Richter. 2010. "Die Regierung des Alterns: Analysen im Spannungsfeld von Diskurs, Dispositiv und Disposition." In *Diskursanalyse meets Gouvernementalitätsforschung: Perspektiven auf das Verhältnis von Subjekt, Sprache, Macht und Wissen,* edited by Johannes Angermüller and Silke van Dyk, 207–236. Frankfurt/Main: Campus Verlag.

Denzin, Norman K. 1978. The research act: a theoretical orientation to sociological methods. New York: McGraw-Hill.

Derrida, Jacques. 1976. *Of grammatology.* Translated by Gayatri C. Spivak. Baltimore, MD: Johns Hopkins University Press.

Derrida, Jacques. 1988. *Limited Inc.,* Translated by Samuel Weber and Jeffrey Mehlman. Vol. 99. Evanston, IL: Northwestern University Press.

Derrida, Jacques, Catherine Porter and Philip Lewis. 1984. "No apocalypse, not now (full speed ahead, seven missiles, seven missives)." *diacritics* 14 (2): 20–31.

Digeser, Peter. 1992. "The fourth face of power." *The Journal of Politics* 54 (04): 977–1007.

Dillon, Michael. 2007. "Governing terror: the state of emergency of biopolitical emergence." *International Political Sociology* 1 (1): 7–28.

Dillon, Michael. 2010. Eventalness: the politics of truth and the analytics of finitude. Unpublished manuscript.

Dillon, Michael and Luis Lobo-Guerrero. 2008. "Biopolitics of security in the 21st century: an introduction." *Review of International Studies* 34 (2): 265–292.

Diprose, Rosalin, Niamh Stephenson, Catherine Mills, Kane Race and Gay Hawkins. 2008. "Governing the future: the paradigm of prudence in political technologies of risk management." *Security Dialogue* 39 (2–3): 267–288.

Elbe, Stefan. 2008. "Risking lives: AIDS, security and three concepts of risk." *Security Dialogue* 39 (2–3): 177–198.

Elster, Jon. 1989. *Nuts and bolts for the social sciences.* Cambridge: CUP.

Ewald, Francois. 1991. "Insurance and risk." In *The Foucault effect: studies in governmentality,* edited by Colin Gordon, 197–210. Chicago: University of Chicago Press.

Floyd, Rita. 2011. "Can securitization theory be used in normative analysis? Towards a just securitization theory." *Security Dialogue* 42 (4–5): 427–439.

Foucault, Michel. 1977. *Discipline & punish.* Random House of Canada.

Foucault, Michel. 1978. *Dispositive der Macht – Über Sexualität, Wissen und Wahrheit.* Berlin: Merve.

Foucault, Michel. 1982. "The subject and power." *Critical inquiry* 8 (4): 777–795.

Foucault, Michel. 1988. *Technologies of the self: a seminar with Michel Foucault.* Edited by L.H. Martin, H. Gutman and P.H. Hutton. London: Tavistock.

Foucault, Michel. 2005. "Die Maschen der Macht." In *Michel Foucault: Schriften in vier Bänden. Dits et Ecrits, Band IV 1980–1988,* edited by Daniel Defer and Ewald Francois, 224–244. Frankfurt/Main: Suhrkamp.

Foucault, Michel. 2007. *Security, territory, population: lectures at the Collège de France, 1977–1978.* Edited by Michel Senellart. Basingstoke, UK: Palgrave Macmillan.

Foucault, Michel. 2008. *The birth of biopolitics:* lectures at the Collège de France 1978–79. Translated by Graham Burchell. Edited by Michel Senellart. Basingstoke, UK: Palgrave Macmillan.

Glaser, Barney G. and Anselm L. Strauss. 1967. *The discovery of grounded theory: strategies for qualitative research.* Berlin: de Gruyter.

Glynos, Jason and David R. Howarth. 2007. *Logics of critical explanation in social and political theory.* London: Routledge.

Gottweis, Herbert. 2003. "Theoretical strategies of poststructuralist policy analysis: towards an analytics of government." In *Deliberative policy analysis. Understanding governance in the network society,* edited by Maarten A. Hajer and Hendrik Wagenaar, 247–265. Cambridge: Cambridge University Press.

Gramsci, Antonio. 1971. *Selections from the prison notebooks of Antonio Gramsci.* Translated by G Nowell-Smith and Q Hoare. New York: International Publishers,.

Guzzini, Stephano. 2011. "Securitization as a causal mechanism." *Security Dialogue* 42 (4–5): 329–341.

Hagmann, Jonas and Myriam Dunn Cavelty. 2012. "National risk registers: security scientism and the propagation of permanent insecurity." *Security Dialogue* 43 (1): 79–96.

Hajer, Maarten A. 1995. *The politics of environmental discourse: ecological modernization and the policy process.* Oxford: Clarendon Press.

Hajer, Maarten A. 2006. "Doing discourse analysis: coalitions, practices, meaning." In *Words matter in policy and planning: discourse theory and method in the social sciences,* edited by Margo van den Brink and Tamara Metze, 65–74. Utrecht: Netherlands Geographical Studies.

Hameiri, Shahar and Florian P Kühn. 2011. "Introduction: risk, risk management and international relations." *International Relations* 25 (3): 275–279.

Hansen, Lene. 2011. "The politics of securitization and the Muhammad cartoon crisis: a post-structuralist perspective." *Security Dialogue* 42 (4–5): 357–369.

Hindess, Barry. 1996. *Discourses of power.* Oxford: Blackwell Publishers.

Howarth, David. 2000. *Discourse, concepts in the social sciences.* Buckinghamshire, England: Open University Press.

Howarth, David. 2010. "Power, discourse, and policy: articulating a hegemony approach to critical policy studies." *Critical Policy Studies* 3 (3–4): 309–335.

Howarth, David and Steven Griggs. 2008. "Metaphor, catachresis and equivalence: the rhetoric of freedom to fly in the struggle over aviation policy in the United Kingdom." In *Political language and metaphor,* edited by Terell Carver and Jernej Pikalo, 197–211. London: Routledge.

Hülsse, Rainer. 2003. "Sprache ist mehr als Argumentation." *Zeitschrift für internationale Beziehungen* 10 (2): 211–246.

Hülsse, Rainer and Alexander Spencer. 2008. "The metaphor of terror: terrorism studies and the constructivist turn." *Security Dialogue* 39 (6): 571–592.

Huysmans, Jef. 2000. "The European Union and the securitization of migration." *Journal of Common Market Studies* 38 (5): 751–777.

Huysmans, Jef. 2006. "International politics of exception: competing visions of international political order between law and politics." *Alternatives: Global, Local, Political* 31 (2): 135–165.

Huysmans, Jef. 2011. "What's in an act? On security speech acts and little security nothings." *Security Dialogue* 42 (4–5): 371–383.

Jackson, Patrick T. 2011. *The conduct of inquiry in international relations: philosophy of science and its implications for the study of world politics.* London, New York: Routledge.

Kavalski, Emilian. 2008. "The complexity of global security governance: an analytical overview." *Global Society* 22 (4): 423–443.

Keller, Rainer. 2006. *Diskursforschung*. Vol. 2. Wiesbaden: VS Verlag für Sozialwissenschaften.

Keller, Rainer. 2008. *Wissenssoziologische Diskursanalyse. Grundlegung eines Forschungsprogramms*. Vol. 58: VS Verlag für Sozialwissenschaften.

Kruse, Jan, Kai Biesel and Christian Schmieder. 2011. *Metaphernanalyse: Ein rekonstruktiver Ansatz*. Wiesbaden: VS Verlag.

Laclau, Ernesto. 1990. *New reflections on the revolution of our time*. London and New York: Verso.

Laclau, Ernesto. 2005. *On populist reason*. London: Verso Books.

Laclau, Ernesto and Chantal Mouffe. 1985. *Hegemony and socialist strategy: towards a radical democratic politics*. London: Verso Books.

Lakoff, George and Mark Johnson. 1980. *Metaphors we live by*. Chicago: University of Chicago Press.

Law, John. 2009. "Actor network theory and material semiotics." In *The new Blackwell companion to social theory*, 141–158. Malden, MA and Oxford: Blackwell.

Lemke, Thomas. 1997. *Eine Kritik der politischen Vernunft: Foucaults Analyse der modernen Gouvernementalität*. Hamburg: Argument-Verlag.

Link, Jürgen. 1984. "Über ein Modell synchroner Systeme von Kollektivsymbolen sowie seine Rolle bei der Diskurs-Konstitution." In *Bewegung und Stillstand in Metaphern und Mythen. Fallstudien zum Verhältnis von elementarem Wissen und Literatur im*, edited by Jürgen Link and Wulf Wülfing, 63–92. Stuttgart: Klett-Cotta.

Lukes, Steven. 1974. *Power: a radical view*. London: Palgrave Macmillan.

Mayer, Maximilian. 2012. "Chaotic climate change and security." *International Political Sociology* 6 (2): 165–185.

McDonald, Matt. 2008. "Securitization and the construction of security." *European Journal of International Relations* 14 (4): 563–587.

Methmann, Chris and Angela Oels. 2015. "From 'fearing' to 'empowering' climate refugees: governing climate-induced migration in the name of resilience." *Security Dialogue* 46 (1): 51–68.

Milliken, Jennifer. 1999. "The study of discourse in international relations: a critique of research and methods." *European Journal of International Relations* 5 (2): 225–254.

Mouffe, Chantal. 2005. *On the political*. London, New York: Routledge.

Neocleous, Mark. 2006. "The problem with normality: taking exception to 'permanent emergency'." *Alternatives: Global, Local, Political* 31 (2): 191–213.

Newman, Saul. 2007. *Power and politics in poststructuralist thought: new theories of the political*. New York: Routledge.

Nonhoff, Martin. 2007. "Politische Diskursanalyse als Hegemonieanalyse." In *Diskurs – Hegemonie – radikale Demokratie: Zum politischen Denken von Ernesto Laclau und Chantal Mouffe*, edited by Martin Nonhoff, 173–193. Bielefeld: Transcript.

Nonhoff, Martin. 2010. "Chantal Mouffe und Ernesto Laclau: Konfliktivität und Dynamik des Politischen." In *Das Politische denken: Zeitgenössische Positionen*, edited by Ulrich Bröckling and Robert Feustel, 33–58. Bielefeld: Transcript.

Nonhoff, Martin. 2011. "Konstruktivistisch-pragmatische Methodik." *Zeitschrift für Internationale Beziehungen* 18 (2): 91–107.

Odysseos, Louiza and Fabio Petito. 2009. "Carl Schmitt." In *Critical theorists and international relations*, edited by Jenny Edkins and Nick Vaughan-Williams, 305–316. New York: Routledge.

Opitz, Sven. 2008. "Zwischen Sicherheitsdispositiven und Securitization." In *Gouvernementalität und Sicherheit: Zeitdiagnostische Beiträge in Anschluss an Michel Foucault*, edited by Patricia Purtschert, Katrin Meyer and Yves Winter, 201–228. Bielefeld: Transcript Verlag.

Patomäki, Heikki. 2015. "Absenting the absence: of future dangers and structural transformations in securitization theory." *International Relations* 29 (1): 128–135.

Pêcheux, Michel. 1984. "Metapher und Interdiskurs." In *Bewegung und Stillstand in Metaphern und Mythen. Fallstudien zum Verhältnis von elementarem Wissen und Literatur im*, edited by Jürgen Link and Wulf Wülfing, 93–99. Stuttgart: Klett-Cotta.

Petersen, Karen Lund. 2011. "Risk analysis – a field within security studies." *European Journal of International Relations.*

Rose, Nikolas. 2001. "The politics of life itself." *Theory, culture and society* 18 (6): 1–30.

Rothe, Delf. 2011. "Managing climate risks or risking a managerial climate? State, security and governance in the international climate regime." *International Relations* 25 (3).

Rothe, Delf. 2012. "Security as a weapon: how cataclysm discourses frame international climate negotiations." In *Climate change, human security and violent conflict*, edited by Jürgen Scheffran, Michael Brzoska, Hans Günter Brauch, Peter M. Link and Janpeter Schilling, 243–258. Berlin/Heidelberg: Springer.

Sager, Sven Frederik. 2000. "System oder Ansammlung. Ist Multimedia überhaupt ein Medium?" In *Sprache und neue Medien: Jahrbuch des Instituts für Deutsche Sprache* edited by Werner Kallmeyer, IX, 406 S. Berlin: de Gruyter.

Sager, Sven Frederik. 2004. *Kommunikationsanalyse und Verhaltensforschung. Grundlagen einer Gesprächsethologie, Stauffenburg-Einführungen; 10.* Tübingen: Stauffenburg-Verl.

Salter, Mark B. 2008a. "Imagining numbers: risk, quantification, and aviation security." *Security Dialogue* 39 (2–3): 243–266.

Salter, Mark B. 2008b. "Securitization and desecuritization: a dramaturgical analysis of the Canadian Air Transport Security Authority." *J Int Relat Dev* 11 (4): 321–349.

Sarasin, Philipp. 2003. *Geschichtswissenschaft und Diskursanalyse.* Frankfurt am Main: Suhrkamp.

Schmitt, Carl. 1993. *Politische Theologie: Vier Kapitel zur Lehre von der Souveränität.* Berlin: Duncker & Humblot.

Schmitt, Carl. 1996. *Der Begriff des Politischen.* Berlin: Duncker & Humblot.

Shannon, Claude E. and Warren Weaver. 1949. *The mathematical theory of communication.* Urbana, IL: University of Illinois Press.

Squire, Vicki. 2013. "Attuning to mess." In *Research methods in critical security studies: an introduction*, edited by Mark B Salter and Can E Mutlu, 37–41. New York: Routledge.

Stritzel, Holger. 2007. "Towards a theory of securitization: Copenhagen and beyond." *European Journal of International Relations* 13 (3): 357–383.

Stritzel, Holger. 2011. "Security, the translation." *Security Dialogue* 42 (4–5): 343–355.

Swyngedouw, E. 2010. "Apocalypse forever? Post-political populism and the spectre of climate change." *Theory Culture and Society* 27 (2–3): 213–232.

Taureck, Rita. 2006. "Securitisation theory – the story so far: theoretical inheritance and what it means to be a post-structural realist." Paper presented at the 4th annual CEEISA convention, University of Tartu, June 25–27, 2006.

Torfing, Jacob. 2005. "Discourse theory: achievements, arguments, and challenges." In *Discourse theory in European politics: identity, policy and governance*, edited by David Howarth and Jacob Torfing, 1–32. New York: Palgrave.

UNGA, United Nations General Assembly. 2009. Sixty-third session, 85th plenary meeting Wednesday, June 3, 2009.

UNSC, United Nations Security Council. 2007a. Sixty-second year, 5663rd meeting, Tuesday, April 17, 2007.

UNSC, United Nations Security Council. 2007b. Sixty-second year, 5663rd meeting, Tuesday, April 17, 2007, Resumption.

UNSC, United Nations Security Council. 2011a. Sixty-sixth year, 6587th meeting, Wednesday, July 20, 2011.

UNSC, United Nations Security Council. 2011b. Sixty-sixth year, 6587th meeting, Wednesday, July 20, 2011, Resumption.

van Munster, Rens. 2005. *Logics of security: the Copenhagen School, risk management and the war on terror.* Faculty of Social Sciences, University of Southern Denmark Odense.

von Lucke, Franziskus, Zehra Wellmann and Thomas Diez. 2014. "What's at stake in securitising climate change? Towards a differentiated approach." *Geopolitics* 19 (4): 857–884.

Vuori, J. A. 2008. "Illocutionary logic and strands of securitization: applying the theory of securitization to the study of non-democratic political orders." *European Journal of International Relations* 14 (1): 65–99.

Wæver, Ole. 1995. "Securitization and desecuritization." In *On security*, edited by Ronnie D. Lipschutz, 46–86. New York: Columbia University Press.

Wæver, Ole 2002. "Identity, communities and foreign policy: discourse analysis as foreign policy theory." In *European integration and national identity: the challenge of the Nordic states*, edited by Lene Hansen and Ole Wæver, 20–49. London/New York: Routledge.

Wæver, Ole. 2004. "Ole Wæver's 10." *Tidsskriftet Politik* 7 (4).

Wæver, Ole. 2011. "Politics, security, theory." *Security Dialogue* 42 (4–5): 465–480.

Wæver, Ole. 2015. "The theory act: responsibility and exactitude as seen from securitization." *International Relations* 29 (1): 121–127.

Walker, Jeremy and Melinda Cooper. 2011. "Genealogies of resilience from systems ecology to the political economy of crisis adaptation." *Security Dialogue* 42 (2): 143–160.

Walter, Jochen and Jan Helmig. 2008. "Discursive metaphor analysis: (de) construction(s) of Europe." In *Political language and metaphor: interpreting and changing the world*, edited by Terell Carver and Jernej Pikalo, 119–131. London, New York: Routledge.

Weber, Max. 1984. *Soziologische grundbegriffe.* Vol. 541. Tübingen: Mohr.

Williams, M. C. 2003. "Words, images, enemies: securitization and international politics." *International Studies Quarterly* 47 (4): 511–531.

Wittgenstein, Ludwig. 2010. *Philosophical investigations.* Oxford: John Wiley & Sons.

Žižek, Slavoj. 1990. "Beyond discourse-analysis." In *New reflections on the revolution of our time*, edited by Ernesto Laclau, 249–260. London: Verso.

Žižek, Slavoj. 2006. "Against the populist temptation." *Critical Inquiry* 32 (2): 551–574.

4 Digging deeper

History and deep structure of climate change and security discourse

The genesis of the securitization of climate change is usually cited as 2007, the year, for example, of the first Security Council debate on climate change and security and the Nobel Prize Committee awarding Al Gore and the IPCC with the Nobel Peace Prize. In this chapter I show that these developments were only the tip of the iceberg – the most visible expressions of a discourse that developed over a longer period and took place in many less visible side stages. Moreover, the issue of climate change and security did not evolve in a vacuum but rather emerged from a range of earlier discourses. In this chapter I will try to dig deeper into the climate security discourse. Digging deeper can first be understood in a temporal sense, as an attempt to trace the historical predecessors and conceptual roots of climate change and security discourse. Second, I also dig deeper into the conceptual deep structure of the discourse in an attempt to carve out the symbolic foundation of articulations of climate security. In the first part of the chapter, I outline the results of a diachronic discourse analysis, reaching back to the early 1990s: genealogical lineages and traces of climate change and security that can be found in discourses on a redefinition of security, environmental security, the security implications of desertification, and the securitization of development policies. In the second part of this chapter, I give a broad overview of the evolution of a genuine climate change and security discourse. The contexts of international politics – or global governance; the Mediterranean region; and the UK – are conceptualized as *three different arenas* in which the discursive struggle on climate change and security is played out. In the final part of the chapter, I present the results of a synchronic, metaphor-based discourse analysis, revealing the deep structure behind recent articulations of climate change and security.

The genealogical roots of the climate security discourse

To understand the securitization of climate change since 2007, one must know the roots of this discourse, which can be traced to the early 1990s (see Detraz and Betsill 2009). A first genealogical ancestor to this discourse is the debate around a redefinition of security.

The 'redefining security' debate

A first genealogical trace of the climate security discourse can be found in the debate about the widening and deepening of security in the aftermath of the Cold War. This thinking gave birth to the call for 'redefining security,' as in the widely disseminated article of the same name by Tuchman Mathews (1989). Besides this academic discourse, a more practically oriented debate has taken place in broader circles of policy advisors, think tanks and security officials. As Stripple notes, the UN played a crucial role in this broader political discourse:

> The 1990s was the decade when the UN, in their search for a post-Cold War direction, launched a range of high-profile meetings on various issues (environment, population, women, housing), and the 1990s saw a tremendous increase in multilateral agreements. This new agenda competed with the traditional security agenda in terms of challenging the apprehension about what the crucial issues of the day were.
>
> (Stripple 2005: 42)

Unlike in academic discourse, the starting point here was not so much a theoretical controversy on the 'appropriate' conception of security but the perceived 'real world changes' in a post-Cold War world. The main threats in the world after the 'end of history' (Fukuyama 1989) seemingly resulted no longer from a direct military confrontation, but from the rise of a variety of transnationally dispersed harms. In Germany as well as in Brussels at the EC/EU level this shift in security policy took place under the label of an *extended security* agenda (Interview with a Berlin-based climate change analyst, Berlin September 14, 2011). Even today this broadened security discourse still constitutes a kind of broader meta-narrative into which the discourse on climate security is embedded, as the following quote from the UNSC debate in 2007 demonstrates:

> National security concerns currently continue to dominate international agendas, while violent conflict, terrorist threats, the danger represented by weapons of mass destruction, the lack of significant progress on disarmament and non-proliferation, and the trade and use of illicit small arms and light weapons certainly remain major challenges to our collective security.
>
> (Terrazas in UNSC 2007b: 29)

Even more important to the current security discourses on climate change was the development of the *human security* concept by the UNDP in this context (UNDP 1994). While in general the UN framework focuses on nation states – that is, the members of the UN – the UNDP report argued for a shift of security policy to the single individual:

> Human security refers to the security of individuals and communities, expressed as both 'freedom from fear' and 'freedom from want.' Severe threats to human security range from genocide and slavery to natural disasters such as hurricanes or floods [...]
>
> (Kaldor *et al.* 2007: 273)

The rationale for this was that some states do not meet their obligations of guaranteeing the safety of their populations and are sometimes even the source of harm. This was informed by the impression of weak or fragile states such as some former Soviet or Soviet satellite states and African countries. A weak central government and the absence of a monopoly on the use of force were seen as major sources of violence that could threaten the wellbeing of citizens (Duffield and Waddell 2006: 6–7).

The environmental security discourse

Besides the more general discourse on a broadened security agenda, the 1990s saw the rise of an academic debate focusing more narrowly on the linkages between environmental issues and security (Barnett 2001; Dalby 2002; Matthew 1995). Some have argued that the thinking on environment and security can be traced back to the eighteenth century, when, for example, the role of glaciers in violent conflicts was taken into account (Phone interview with an Oslo-based climate security researcher, September 29, 2011). While such thinking remained within the realm of science for the next 200 years, this changed with the publication *Limits of Growth* by the Club of Rome in 1972 (Meadows *et al.* 1972). The bleak scenarios painted in this report, two of which anticipated an overshoot and collapse of human and earth systems in the twenty-first century, have strongly influenced subsequent scientific inquiries that investigated the links between natural resource scarcity and violent conflicts. A similar politicizing effect was created by the 1987 so-called Brundtland report *Our Common Future* (Brundtland Commission 1987) by the World Commission on Environment and Development (Interview with a climate security analyst, Berlin September 14, 2011). In addition to creating broad public awareness, this report brought together the issues of development and environmental change, and gave prominence to the concept of sustainability as a solution to the global environmental problems outlined by the Club of Rome and others.

In the 1990s, several groups of conflict researchers started investigating the causal links between environmental degradation and the likelihood of violent conflicts in developing countries (see Hartmann 2010: 236). The most attention was received by the works of the so called Toronto group.[1] In a nutshell, the Toronto group, led by Thomas Homer-Dixon, tried to empirically test neo-Malthusian arguments that linked resource scarcity and population growth to the possibility of violent conflict on the basis of qualitative case studies (Homer-Dixon 1991). With their first research project the group studied the effects of different types of environmental changes on violent conflict: reduction of the stratospheric ozone layer, land degradation, deforestation, water degradation, depletion of global fish stocks and climate change (Homer-Dixon 1994). The overall insight produced by this research was that none of these factors alone could explain the occurrence of violent conflicts. However, environmentally induced resource scarcity in combination with other socio-political factors such as political instability indeed could cause distributional conflicts or exacerbate existing conflicts.

However, there has also been a different line of thought that has informed the debate on environmental security. This discursive strand argued that environmental challenges would fundamentally affect the ways we ought to think about security (Barnett 2001; Dalby 2002). This refers to the referent objects of security: threatened by environmental problems are individuals on the one hand and the whole planet, nature or living species on the other. In addition, it also means that the threats differ from a conventional understanding of security as these are non-violent phenomena, i.e. threats without a subject status. We will come to this important point below. Lastly, responses to these novel threats have to change, according to this understanding of environmental security (Schwartz 2003: 3).

For the present analysis, it is important to note that the climate change and security discourse grew genealogically out of the broader environmental security discourse (Phone interview with a member of the European External Action Service, November 25, 2011). The continued existence of core storylines and concepts illustrates this clearly. By example, in 1991 Thomas Homer-Dixon wrote that:

> [...] warmer temperatures could lead to *contention over new ice-free sea-lanes in the Arctic* or more accessible resources in the Antarctic. Bulging populations and land stress may produce *waves of environmental refugees that spill across borders* with destabilizing effects on the recipient's domestic order and on *international stability*. Countries may fight over *dwindling supplies of water* and the effects of upstream pollution. In developing countries, a sharp drop in food crop production could lead to internal strife across urban–rural and *nomadic–sedentary cleavages*.
>
> (Homer-Dixon 1991: 77, emphasis added)

The arguments that I have emphasized in this statement from over 20 years ago could easily have come from the current climate security discourse (see Floyd 2008: 60–61). That being said, it is important to note that climate change was not a major issue in the environmental security debate but was rather generally taken as one amongst several global environmental challenges. Furthermore, research such as Homer-Dixon's and its empirical basis became fiercely disputed in the following years

(Phone interview with a UK climate security analyst, December 14, 2012). The most prominent criticism of the early environmental conflicts literature was articulated by a group of scholars at the International Peace Research Institute Oslo led by Nils Peter Gleditsch (Gleditsch 1998). The scholars criticized Homer-Dixon and his followers for the inherent neo-Malthusianism and its underlying narrative, which Gleditsch describes as:

> Population growth/high resource consumption per capita → deteriorated environmental conditions → increasing resource scarcity → harsher resource competition → greater risk of violence.
>
> (Gleditsch 1998: 838)

At about the same time, different scholars opposed this general storyline of the resource scarcity literature with an antithetical hypothesis – namely, that abundance of valuable natural resources would stimulate conflicts. This argument gave birth to the so-called greed-versus-grievance debate (see e.g. Collier and Hoeffler 2005; De Soysa 2002).[2]

The scientific uncertainties meant that the discourse remained mainly in the realm of academia:

> [...] you saw until the mid-1990s a lot of writing around the environment, security, forest resources, oil access, water access. But then that died down somewhat, not in terms of the number of papers that were written but in terms of the attention they were given until the early 2000s.
>
> (Phone interview with a UK security analyst, December 14, 2011)

This situation, however, changed in the early 2000s with increased attention given to the threat of climate change.

The Mediterranean and desertification as a security threat

While the conceptual traces that have been outlined so far apply to the climate security discourse in general, independent from its spatial manifestations, there is a particular genealogical source for the climate security discourse on the Mediterranean: the securitization of *desertification, and soil and land degradation* (DSLD) (see e.g. Brauch and Spring 2009). The discourse on desertification was not restricted to the academic realm; it had considerable resonance in various policy circles.[3] One can explain the prominence of storylines linking desertification and security by the growing problematization of seemingly *ungoverned spaces* in the aftermath of 9/11. Ungoverned spaces, such as, for example, deserts, can hardly be controlled by state forces and are hence a safe haven for radicals and terrorist movements (Mitchell 2010: 289–290).[4] What is more, desertification discourse resonates well with older colonial stereotypes that problematize overgrazing and population growth as root causes of social problems in North African countries (Hartmann 2006: 234). Last, southern European governments, such as that of Spain, saw themselves confronted with a threat of desertification on their own territories. Hence they perceived it as a problem shared with southern Mediterranean countries.[5] The problematization of desertification was not restricted to European actors but was also shared by positions from within Middle Eastern and North African (MENA) countries. For example, at the 2003 NATO workshop "Desertification in the Mediterranean region: A security issue?" (Kepner and Rubio 2003), there was widespread participation of officials from southern Mediterranean countries.

The discourse on the security implications of desertification took place in the context of the general environmental security debate that gained momentum in the 1990s. Out of the various environmental threats that have been discussed with respect to the region, DSLD was given special attention, which showed up in a number of high-level workshops, reports and assessments in the early 2000s. Initially, therefore, it was not climate change but rather desertification that became the central *nodal point* that structured the discourse on environment and security in the region. This is illustrated in the following articulation:

> There is strong evidence that desertification exacerbates poverty and negatively affects social order and stability. It can contribute significantly to water scarcity, famine, internal displacement of people, migration, and social breakdown and thus presents a recipe for political, social, and economic instability which can also lead to tension between neighboring countries and armed conflict.
>
> (Kepner and Rubio 2003)

What is fascinating about this quote is that one could – without any loss of meaning – substitute *desertification* with *climate change* and would immediately find oneself in the familiar domain of the climate security discourse of the late 2000s. Further similarities between the discourses on desertification and climate change exist with respect to major storylines. For example, desertification has been prominently linked to the problem of environmental migration from as early as 1994: "It is estimated that over 135 million people may be at risk of being displaced as consequence of severe desertification" (Westing 1994: 180). In 2006, the UN University released a study entitled *Overcoming one of the greatest environmental challenges of our times* (Adeel *et al.* 2006). Like climate change, DSLD is framed in metaphors of struggle and survival, as exemplified in the following: "Desertification and drought are slow – but effective – killers that do not grab headlines like tsunamis or earthquakes" (Kalbermatten 2009: 22). The point here is not to argue that desertification and climate change security discourses would be interchangeable. Rather, climate security discourse on the Mediterranean drew on particular storylines and rationales from the securitization of DSLD and rearticulated them in terms of climate security.[6]

The national level and the development–security nexus

In the context of Western nation states such as the UK, the concrete manifestation of the climate change and security discourse cannot be understood if one does not consider the securitization of development that followed 9/11. The initial response of the US and its allies to the terrorist attacks of September 11, 2001 followed a rather geopolitical security script, as the military punitive expeditions in Afghanistan 2001 and Iraq 2003 demonstrated. At the same time, however, the dislocative moment of 9/11 challenged security discourses in a way that paved the way for a fundamental rethinking of security (De Goede 2008: 163–165; Duffield and Waddell 2006: 10–12). This is exemplified by the following quote from the National Commission on Terrorist Attacks Upon the United States, also known as the 9–11 Commission: "In the post-9/11 world, threats are defined more by the fault lines within societies than by the territorial boundaries between them. From terrorism to global disease or environmental degradation, the challenges have become transnational rather than international" (9–11 Commission 2004: 361–362). Hand-in-hand with this shift in the conceptualization of security went a rethinking of the terrorism threat through the prism of what can be labeled a "root cause theory of terrorism" (Schwartz 2003: 4). As its name already suggests, the root cause theory suggests that the sources of terrorism are to be found in the socio-economic conditions under which people live: "terrorism in not born but bred" (Schwartz 2003: 4). From this simple assumption, it is only a small step to establish the link between environmental degradation and the spread of terrorism: as the living conditions of people deteriorate, they become more open to radical ideologies and movements. Furthermore, environmental problems can deepen existing social antagonisms within societies. Last, they further undermine the governmental capacities to resolve these existing tensions (Schwartz 2003: 9). Several authors from the environmental conflicts literature have established this link between the environment and terrorism (Homer-Dixon 1999: 3; Klare 2007).

The growing attention to the socio-economic root causes of terrorism and political extremism led to an increasing securitization of development (Abrahamsen 2004). For example, US president Bush stated at the Monterrey Development Summit 2002 that "we fight against poverty because hope is an answer to terror" (Bush quoted from Piazza 2006: 160). Informed by the notion of living in an increasingly networked world, it was feared that poverty-related grievances could not be separated from the security of Western nation states (Duffield and Waddell 2006: 43). This is perfectly expressed in the following statement by then UK foreign minister Jack Straw: "It is no longer necessary to prove a direct link between a troubled faraway country and the order of our own societies" (Straw quoted from Abrahamsen 2005: 65). Helping the poor to develop, therefore, was taken out of a mere economic or normative context and presented as a national interest. A similar storyline that gained momentum at that time was that without development there would be no security, and without security there would be no development (Kühn 2010). Aid and official development assistance (ODA) were increasingly seen as a means to achieve stability and to prevent violent conflict in fragile or conflict-prone states and regions.

Another manifestation of the development–security nexus can be found in the growing importance of failed states in Western security discourses and practices (Helman and Ratner 1992). The metaphor of failing or failed states is generally used to describe nation states that "cannot or will not safeguard minimal civil conditions for their populations: domestic peace, law and order, and good governance" (Jackson 2000: 296). The international community began to be concerned with failing or fragile states and possible state collapses in the aftermath of the Cold War. Failing states were depicted as risk factors for international security by threatening neighbors through refugee flows and political instability (Helman and Ratner 1992). In the early 2000s following the 9/11 terrorist attacks, we notice that the threat of failed states slowly replaced the metaphor of rogue states as the emblematic enemy figure in Western security discourses. The reasons for this shift after 9/11 seem obvious: "Failed states have become to be feared as "breeding grounds of instability, mass migration and murder" (in the words of Stephen Walt) as well as reservoirs and exporters of terror'" (Rotberg 2002: 128). While there is not *the one* single policy that coheres with such a development–security nexus, it is fair to say that the growing focus on state building since the early 2000s, and with it a growing number of civil–military cooperation projects, goes mainly back to the growing prominence of the discourse outlined in this current section (Rotberg 2002: 128).

The discursive interplay between security and development represents an important genealogical root of climate change and security discourse. First, the central storyline that instability elsewhere directly impacts security in Western states is one of the major arguments in the climate security discourse, as we will see in the next chapters. Additionally, the dual motives of normative responsibility and national interest to provide assistance are later found in the securitization of climate change impacts, especially in the adaptation discourse. Last, the development–security nexus also promoted a couple of political concepts – such as vulnerability, human security, and resilience – that also became important in the securitization of climate change. Yet such concepts did not spill into the climate security debate in a unidirectional transfer. Rather, one has to understand these different discourses as a complex interplay between various specialized discourses. They mutually imbricate each other and exchange rationalities

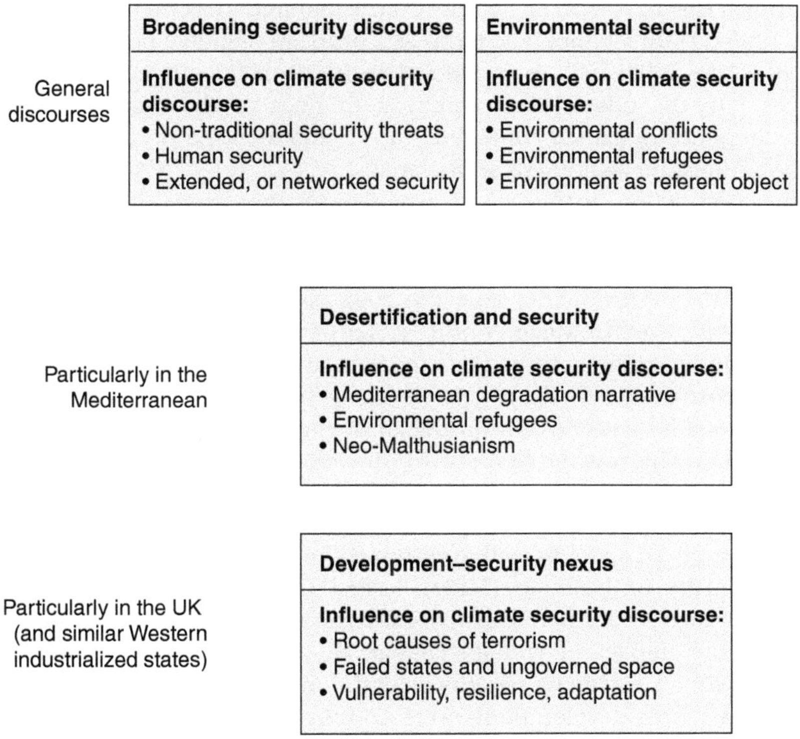

Figure 4.1 The genealogical roots of climate security discourse (source: author).

and storylines as well as concepts or framings (Chandler 2007: 364; Stern 2010: 10–11).

Figure 4.1 summarizes the different conceptual roots of climate change and security thinking as well as their impact on climate security discourse.

Emergence of climate security as a matter of concern

Climate change was initially related to security concerns in the context of an engagement with environmental security issues in general. At times, the rationales made in this academic discourse found their way into broader political contexts. For example, Peter Gleick (1989) published an article on the implications of climate change for international security that featured also in the ECO newsletter published by the Climate Action Network, the largest NGO network in the field of global climate governance (Gleick 1990). Such expressions, however, remained sporadic; there was nothing like a genuine climate security discourse in the sense of an ongoing rearticulation of such arguments.[7] On the one hand, this goes back to the dominant articulations of climate change at that time: different studies have convincingly shown that the dominant framings of climate change in the 1990s understood it in terms of a market failure or an economic inefficiency problem (Hajer 1995: 31), in technocratic terms as a problem of global political management (Lutes 1998: 158), or as a global justice problem (Adger *et al.* 2001: 699–700). On the other hand, empirical findings of the environmental security literature remained contested – not to mention the fiercely contentious global warming hypothesis at the time.[8]

In the early 2000s, further single articulations that explicitly dealt with the relationship between climate change, conflicts and security can be observed (see Trombetta 2012: 156–157). In Germany, the Ministry of Environment commissioned a report in 2002 asking the question whether climate change would induce conflict potentials (Brauch 2002; Carius *et al.* 2002). One year later, futurists Peter Schwartz and Doug Randall produced a report titled *An Abrupt Climate Change Scenario and Its Implications for United States National Security* on behalf of the Pentagon (Schwartz and Randall 2003). This was the first report on climate security commissioned explicitly by a security actor. Although it focused on security implications for the United States, the Pentagon report received major attention in other industrialized states as well. At around the same time, in 2003–2004, the UK Ministry of Defence (MoD) "asked the meteorological office, which is a branch under the Ministry of Defence [...] to do some strategic studies on the impact of climate change in their key strategic areas'" (Interview with a UK climate security analyst, London October 20, 2011).

These discursive developments were supported by the occurrence of particular extreme weather events or natural disasters, including Hurricane Katrina, which hit New Orleans in 2005, and the summer heat wave in Europe in 2003. What is important is not the occurrence of such events per se but the fact that these have been increasingly articulated as an expression or indicator of climate change. For example, simultaneous to the 2003 heat wave, the World Health Organization published a report on the implications of climate change on global health that argued that climate change was already causing excess death at the present time (Phone interview with a UK climate security researcher, November 2, 2011). Hurricane Katrina, on the other hand, was not only often linked to the phenomenon of climate change, but was moreover articulated in terms of a national security crisis:

> The fact that [...] even the military had to intervene to avoid a crisis and to restore the public order of a whole state, I believe, that was a shock for the US and probably also for the rest of the world.
>
> (Interview with a German government advisor, Berlin September 14, 2011, author's translation)

Furthermore, the event of Katrina was articulated in relation to the issue of climate-induced migration: "The displacement of thousands of US Gulf

Coast residents in the aftermath of Hurricane Katrina is emblematic of a human migration challenge that will likely become more severe in the years and decades ahead" (Smith 2007: 617).

Another important date in the development of a climate change and security discourse was the publication of the *Economics of Climate Change* report by economist Sir Nicholas Stern (Stern 2006). "Nick Stern didn't really look at the geopolitics of it, but the Stern Review repeated the view that climate change could cause conflict, could cause international migration, on two hundred million environmental refugees by 2015" (Phone interview with a UK climate security researcher, November 2, 2011). In the same year, the documentary film *An Inconvenient Truth* by Al Gore attracted international attention: "I think Al Gore did a very powerful job in disseminating the arguments" (Phone interview with a member of DG Humanitarian Aid, September 27, 2011).[9] In early 2007, then, the results of the fourth IPCC report slowly became public and attracted hitherto unprecedented attention. It presented the climate science results with a certainty previously unseen and presented a dark picture of regionally differentiated climate change impacts (IPCC 2007).[10] The development of climate security discourse and the critical discourse moments identified in this section are summarized in Figure 4.2.

How climate change and security entered global governance

Together the outlined developments meant that climate change and security articulations produced increasingly greater resonance in industrialized

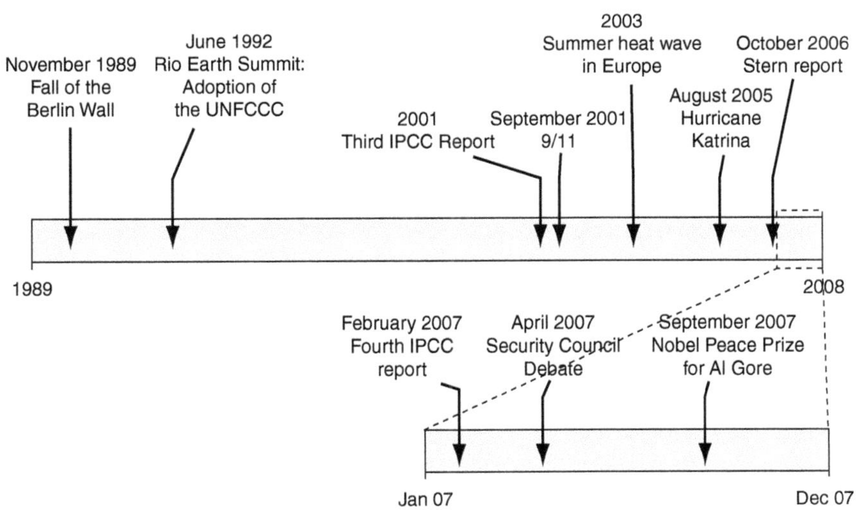

Figure 4.2 Critical discourse moments in the climate change and security discourse (source: author).

states.[11] The main stage for the evolving discourse at the international level was the governance structure of the UN. It is important to consider the particular governance structure of the UN, split between open and inclusive bodies such as the General Assembly and the UNFCCC, and the exclusive and restricted Security Council. While the main body for the debate on climate change related issues has been the UNFCCC since 1995, the discourse on climate change and security within the UN framework has mainly been staged in the UNSC and the UNGA. In November 2006, the soon-to-be UN Secretary-General Ban Ki-moon called climate change the world's greatest security threat, compared it to the destructive potential of the Second World War and claimed that it had – among other sources – caused the Darfur conflict (Ki-moon 2007). Subsequently, Great Britain brought the issue of climate change and security into the Security Council for the first time ever in April 2007. The debate included statements from over 50 delegations from the five permanent members, non-permanent members such as Germany and India, and also from particularly threatened countries such as some Least Developed Countries and Small Island Development States (SIDS) (UNSC 2007a).

A fierce antagonism surfaced in the first climate security debate in the UNSC between the European 'forerunners' that tried to push for a more active role for the UNSC in international climate governance and the group of developing countries (G77), which saw the maneuver by the UK and other European countries as an attempt to hijack the climate issue from inclusive bodies such as the UNFCCC (see Sethi 2011). Countries such as China and India, which sought to establish themselves as leaders of this opposition camp, thus vehemently defended the idea that climate change first and foremost presents a development issue and not a security issue (e.g. Sen in UNGA 2009b: 21–23). As Russia also remained skeptical towards the linking of security and climate change, the Security Council debate in 2007 ended without making any policy proposals.

The 2009 General Assembly debate on the security implications of climate change took a somewhat different turn because it was initiated by the group of SIDS, which portrayed climate change as a threat to their very national survival (e.g. Moses in UNGA 2009b: 1–3). This transcended the existing fissure between Western proponents of climate security arguments and the G77 camp, as no participant could reasonably question the threat to the SIDS. This had the effect of adding an ethical and normative tone to the discourse. Nevertheless, the SIDS proposal resulted in a fierce debate in which G77 countries repeated their skepticism of such a security framing of climate change (see MacFarquhar 2009). Due to the mediation efforts of Indonesia, a consensus declaration could be adopted that especially highlighted climate change-induced existential threats for SIDS and to human security in vulnerable regions (UNGA 2009b). It asked the Secretary-General to prepare an overview report on the security implications of climate change that would bring together the viewpoints and experiences of all member states in a comprehensive manner. This report was published in 2009 (UNGA 2009a). Under the German presidency, the Security Council in July 2011 took up the issue of climate change and security a second time. Although resentment of the G77 and non-aligned movement participants was still present, the debate brought about a joint statement. The statement acknowledged potential international security implications of climate change in the future (Callick 2011).

The debates in the major UN bodies certainly represented the most prominent articulations of climate change and security discourse in the international arena. However, there were also several *parallel sites* of discursive struggle at the international level involving more specialized UN bodies such as UNDP, or UNEP. While these organizations have not contributed to the discourse through authorized speech acts,[12] they contributed to spreading the discourse through workshops and programs on particular problems, often with a regional focus.[13] The International Organization for Migration (IOM) is one of the key players in global governance when it comes to the issue of climate-induced migration (CIM). The IOM runs a comprehensive research program on migration, climate change and the environment, has induced policy dialogues on CIM at different regional levels, and has even brought the issue of CIM into the UNFCCC process (IOM 2009).

NATO has introduced a chapter on environment and security (§15) into their latest security doctrine adopted in November 2010. While NATO has no formal mandate to engage in climate change and security in particular, it participates in the discourse in several ways.[14] Since August 2010 the Emerging Security Challenges Division has worked conceptually on non-traditional security threats such as climate change. In addition, civil emergency and disaster relief activities are increasingly framed against the backdrop of climate change.[15] Certainly the most important supranational organization with respect to the climate change and security discourse is the EU. Following a joint report of the EU High Representative Javier Solana and the European Commission (Solana 2008), the EU adopted the "climate change and international security" process (Council 2009).

How the Mediterranean became seen as a hotspot region

The discourse around the security implications of climate change in the Mediterranean – or MENA (Middle East and North Africa) region – is staged largely within the same institutional setting. The reason is that it is mostly an 'exported' discourse with interventions coming mainly from either international organizations or scientists and think tanks in northern industrialized countries.

As I described regarding the international arena, there has been a longer history of environmental security concerns in the region in general.

A good summary of this early discourse is found in an edited volume on *Security and Environment in the Mediterranean* (Brauch *et al.* 2003). This discourse from the very beginning also comprised articulations from southern (Mediterranean) countries (Interview with a climate security researcher, Hamburg, November 17, 2011). The broader environmental security discourse on the region painted the picture of an endangered MENA region particularly challenged by issues such as population growth, water scarcity and desertification. The impacts of climate change up until the late 2000s, however, were subsumed under other environmental problems. Georgas, for example, discusses climate change impacts on Mediterranean coastal areas under the rubric of population growth (Georgas 2003).

When the discourse on climate change and security became internationally prominent, nearly the same issues that had been discussed with respect to environmental security were adopted and gathered under the common umbrella of climate change and security. Although it did not explicitly deal with the Mediterranean,[16] the IPCC special report on the regional impacts of climate change in 1997 was a critical juncture in this respect, as it included for the first time a differentiated assessment of regional climate change impacts (Watson *et al.* 1997). As a result, nearly every study or report of Western NGOs and think tanks on climate change and security included a special focus on the MENA region – or, at least, parts of it such as the Middle East or North Africa (Campbell *et al.* 2007; Center for Naval Analysis 2007; WBGU 2009). The climate change and security discourse on the region connects with already existing security discourses, such as those on migration, political Islam and terrorism. Hence, it comes as no surprise that security actors and security-related think tanks showed a particular interest in the region (NIC 2009). The Organization for Security and Co-operation in Europe (OSCE) plays an important role and maintains special relationships with several Mediterranean countries (Maas and Tänzler 2009; Tänzler *et al.* 2010). As a regional security organization, the OSCE displays a strong sensitivity to regional security challenges such as the one climate change poses for the Mediterranean. Major international organizations have also been influential in spreading the discourse in the region. The World Bank, for example, has had a particular focus on climate-induced migration in the region.[17] In addition, many interviewees stressed the importance of UNEP: "The one actor which strikes me as the strongest one is UNEP, which is in Nairobi, which has a more or less explicit agenda on pushing this kind of issue forward" (Phone interview with a climate security researcher, September 29, 2011).

None of the actors mentioned so far, however, explicitly mentions *the Mediterranean* region as a referent object – rather it is referred to as North Africa, or the Middle East, or MENA as a combination of both. This differs from the European Union: the European Security Strategy (ESS) formulated in 2003 dedicated a whole section to its unstable (southern) neighborhood (EU Security Strategy 2003: 7–8). While the ESS did not explicitly mention climate change as a security concern,[18] this changed with the adoption of the EU's climate change and international security (CCIS) process in 2008 (see Solana 2008). In the paper *Climate Change and International Security*, the Commission and the High Representative explicitly articulated their concern for climate-related threats stemming from Europe's immediate neighborhood:

> Since the EU's neighbours include some of the most vulnerable regions to climate change, e.g. North Africa and the Middle East, migratory pressure at the European Union's borders and political instability and conflicts could increase in the future. This could also have a significant impact on Europe's energy supply routes.
>
> (Solana 2008: 6)

All interviewees agreed that, up until now, the climate change and security discourse had not produced much resonance within the southern Mediterranean countries (e.g. Phone interview with a climate change and energy activist, September 26, 2011).[19] There are two main reasons for the absence of discourses on climate change and security. First, public attention to the issue of climate change is generally low in many Arab countries (Eskjaer 2009).[20] Brauch (2012: 751), for example, noted a low willingness of such countries to participate in UNFCCC efforts. Out of the five North African countries, Algeria, Egypt and Morocco have to date submitted two of five national communications on national CO_2 emissions and mitigation and adaptation measures that the UNFCCC obliges participating states to publish, while Tunisia has submitted one and Libya, so far, none. The second reason for the minor significance of the climate change and security discourse in Arab countries is that "[...] concepts of 'softness' and 'security' do not match in Arab thinking as the Arab region is plagued by mostly military and territorial (hard) security issues" (El-Sayed Selim 2011: 2011). Moreover, as one interviewee noted, in the authoritarian MENA regimes "everything that counted for political elites was regime security and everything else was largely irrelevant" (Interview with a climate security researcher, Hamburg, November 17, 2011).

However, a few exceptions exist. For example, single southern Mediterranean scientists have been dealing with the security implications of environmental change (Interview with a climate security researcher, Hamburg, November 17, 2011). Furthermore, single NGOs such as the Lebanese IndyAct or the Friends of the Earth Middle East have partly articulated arguments about climate change and security in the region (Phone interview with a regional climate change activist, September 16, 2011; phone interview with a Geneva-based environmental security analyst, September 15, 2011). Last, regional divisions of international organizations such as the UNDP's Regional Bureau for Arab States have also participated in the discourse on climate change and security in the region

(Elasha 2010: 25–27). All of these articulations, however, come from actors that are marked by their Western orientation, as such employees of international relations or scientists/activists have strong connections to Europe because of scholarly exchange and visits abroad. To sum up, one could thus conclude that an emerging *transatlantic discourse coalition* on the climate security implications in the region was occasionally supplemented by single articulations from MENA countries (see Chapter 6).

How the UK became a forerunner on climate security

The widest discursive dissemination and the clearest institutional effects of a securitization of climate change can be observed in the UK (Interview with a German governmental advisor, Berlin September 14, 2011). It was, after all, the initiative of the UK's Foreign and Commonwealth Office (FCO) that put the topic on the agenda of the UN Security Council. However, as early as the late 1980s, the then Prime Minister Margaret Thatcher framed climate change as the main risk to humanity (Carvalho and Burgess 2005: 1463). For example, in 1989 at the UN, Thatcher argued that "it is life itself that we must battle to preserve," and that "the challenge for our negotiators is as great as for any disarmament treaty" (Thatcher quoted from Carvalho and Burgess 2005: 1463). In the 1990s, however, public and political attention to the issue of climate change decreased, and so did articulations of climate change and security linkages. A crucial dislocative moment for the UK was the severe flooding in 2000, which was called a 'wake-up-call' for then Deputy Prime Minister John Prescott (see Rogers-Hayden *et al.* 2011: 136). In 2003–2004 the British MoD started assessing the security implications of climate change for the UK. At the same time, the Development, Concepts and Doctrine Centre (DCDC), a think tank and training center of the British military, commissioned a range of projects on the security implications of climate change:

> They did a – and this must have been in 2005 when the project came out, and in 2006, they did a long term 2030 trends document, in which both climate change and resources figured quite high in the military planning process. That was probably the first major public document issued by a ministry of defence, which had a strong theme on that.
>
> (Interview with a UK climate security analyst, London October 21, 2012)

In addition, the FCO made climate change and the international promotion of climate change issues one of its priorities. In 2004 it established a climate change and energy department (Boas 2015: 87). It has been argued that the UK tried to re-establish itself as a major international power by becoming a forerunner in global climate politics. It was then foreign minister Margaret Beckett who, in 2006, promoted the storyline of climate change as a threat to international security in UK public discourses. In 2009 this led to the creation of a working group in the FCO tasked explicitly with the issue of climate and energy security (Boas 2015: 91). Next to these governmental actors were several UK think tanks that started working on the issue of climate change and security. Among them was the London-based organization Third Generation Environmentalism – E3G (e.g. Mabey 2008). Since its foundation in 2004, E3G has worked on the issues of and linkages between climate change, energy and security (Interview with a UK climate security analyst, London October 21, 2011). Other high-profile think tanks and NGOs working in the field of security, such as the *Royal United Services Institute* (RUSI) (Phone interview with a UK climate security analyst, December 14, 2011) or International Alert (Smith and Vivekananda 2007) followed.

The publication of the UK's first national security strategy in 2008 marked another crucial juncture in the development of UK climate change and security discourse: "It was really a landmark moment for us in terms of how we took climate security forward and really where we are today" (Phone interview with a member of the UK FCO, December 15, 2011). Climate change was now being referred to as "potentially the greatest challenge to global stability and security, and therefore to national security" (Cabinet Office 2008: 18) in an official defense document by the government of one of the world's most important industrialized countries. Although it does not explicitly deal with the security implications of climate change, the latest national security strategy update as well as the strategic security and defense review (both 2010) were also clearly influenced by this discourse: "the whole aim was to deliver some sort of vision and recognize the threats that were potentially on the horizon if we didn't do something about it." (Phone interview with a member of the UK FCO, December 15, 2011). Further critical discourse moments include the publications of the two Foresight reports on *The International Dimensions of Climate Change* (Government Office for Science 2011b) and *Migration and Global Environmental Change* (Government Office for Science 2011a) in 2011 (Phone interview with a member of the UK FCO, December 15, 2011). These reports are particularly important as they are indicative of a shift from climate security storylines and the promotion of drastic mitigation measures towards a strategy of resilience and a focus on adaptation (Methmann and Oels 2015).

Deep structure of the climate change and security discourse

In 2003, the then head of the UK Meteorological Office and ex-IPCC head Sir John Houghton stated that climate change would be a weapon of mass destruction, killing more people than terrorism (Houghton 2003). This statement produced considerable resonance in the UK and was mentioned by different interviewees as the kick-off to the discourse on climate change and security in the UK (Phone interview with a UK climate change and security researcher, November 2, 2011). The statement in the following years circulated through the discourse as a quote, creatively adopted and reformulated for example in the expression "weather of mass destruction" (Brown 2007). The considerable effect of Houghton's statement can partly be explained by the particular context of the articulation: "That was in early 2003, which in the context of UN weapons inspectors looking for weapons of mass destruction in Iraq, and Britain being part of the coalition about to invade Iraq, was a politicized thing to say" (Phone interview with a UK climate change and security researcher, November 2, 2011). However, the wide dissemination of this metaphorical statement also rests upon the fact that Houghton draws on and 'exploits' a particular discursive deep structure – war metaphors – that is a prominent feature of our interdiscursive vocabulary. Hence, Hougthon's articulation was not a singularity, but is embedded within countless articulations that metaphorically frame climate change in terms of war and battle. In this section I set out to reveal the collective symbols that mark the discursive field upon which climate change and security storylines are constructed.

The field of the interdiscourse provides a set of symbols, metaphors and analogies that make it possible to grasp and talk about the abstract and complex phenomenon of climate change.[21] In the present section I outline the basic collective symbols that together constitute this deep structure and reveal their political implications. As the discursive deep structure is the same in the three manifestations of the climate security discourse studied, it is presented prior to the single case studies. The most basic conceptual metaphors that could be identified in the empirical material are *ontological metaphors* (Lakoff and Johnson 1980: 25–28). Those are grounded in most basal human experiences with objects of the real world. When it comes to physical or social processes that cannot be directly perceived, ontological metaphors are necessary to render them thinkable and speakable. To be able to grasp climate change, it is discursively translated into an object, a person, or a substance/material. Though this might sound trivial at first, it is nevertheless important to note, since ontological metaphors are a necessary condition for drawing causalities or attributing climate change: "the climate is the hull, in which humans live" (Spiegel Online 2009), author's translation); "The planet may not wait patiently" (Meeks and Shank 2008).

The anthropomorphization of climate change enables one to attribute certain negative characteristics to it or place responsibility for certain circumstances on it. Thus, ontological metaphors serve as a starting point for further specification of the threat of climate change.

Metaphors of war and struggle

The most prominent collective symbols in the climate security discourse revolve around metaphors of *war and struggle*. "We are confronted with a chemical war of immense proportions" (Weisleder in UNSC 2007b: 32); "the warming war" (Pita in UNSC 2007b: 8); "the fight against climate change" (Morgan in UNSC 2011b: 9). The war metaphor of climate discourse comprises several sub-concepts such as climate change as an *enemy*, a *weapon*, a *criminal*, a *threat* or a *catastrophe*. A trope closely related to the war metaphors in the climate security discourse is the personification of climate change. The discursive effect is that climate change appears as a clearly identifiable, hostile subject with the capability to act: "climate change can further erode state capacity" (Rice in UNSC 2011a: 7); "we are at the mercy of inclement climatic variations that hinder the sowing of and destroy our crops" (Flores in UNSC 2011b: 12). Through this anthropomorphization, it becomes possible to articulate climate change as an external enemy in security storylines. Furthermore, it allows climate change to be further specified and predicated with certain human attitudes such as being *unfair, cruel, grim, erratic* or *dangerous*. Climate change is thus not portrayed in a neutral way but, rather, negatively predicated with characteristics usually attributed to potential aggressors. This indicates one prominent narrative script in the discourse, stating that "nature rebels against humans" (Flores in UNSC 2011b: 12).

As climate change is constructed as an enemy or threat, it logically follows that there needs to be defenders as well. This corresponds to the oft-used concept that *climate politics is a struggle or war*. In the empirical material, two different dimensions of the metaphorical frame can be identified. The first dimension constructs *humanity* or the *international community* as one party in the *struggle against the enemy climate change*, while the second dimension paints a picture of a *struggle between different countries* over their responsibility for global warming and their contribution to climate politics. It is important to note, though, that the existence of war metaphors alone does not necessarily mean a securitization of climate change. Metaphors of battle and war can be found in different contexts, for example in economic discourses (see Koller 2004). Nevertheless, as one could see from the examples, war metaphors and collective symbols convey certain irreducible implications: first, they articulate an antagonistic relationship – between an enemy/threat and something threatened, and second, they produce a sense of urgency. The nature of the threat as well as the policy measures that are brought forward, however, depend on the storylines and narratives in the specific discourse.

Topological symbols and motional metaphors

Related to collective symbols of war and struggle are concepts around the notion of contestation, as expressed in the common statement: "climate change is a challenge."[22] Further related concepts draw on these contest metaphors and link them with motional collective symbols to make sense of the temporal dimension of climate change. One such symbol is the metaphor of *climate politics as a journey*, which rewrites political progress (or time) in spatial terms. In line with those motional concepts, *climate change*, or *the warming earth, is a vehicle*, moving at an ever-faster rate towards a danger zone: "Climate change [...] is accelerating in a dangerous manner [...]. [W]e were on the cusp of a global climate disaster without an urgent and ambitious collective response" (Goddard in UNSC 2011b: 228–229). This is logically consistent with the prominent metaphorical concept that *time is a landscape we move through* (Lakoff *et al.* 1991: 77). These motional symbols are often further specified through metaphors of a race, contestation, or run: "today's debate will jump-start our search for wise and insightful solutions" (Kim Sook in UNSC 2011b: 6); "The Council needs to keep pace with the emerging threats of the twenty-first century" (Rice in UNSC 2011a: 7); "the international community needs to accelerate its efforts" (Çorman in UNSC 2011b: 30). The earth, its nation states, and, indeed, the whole of humanity are constructed as vehicles in a race or contest against one another. Such collective symbols are used to highlight the *urgency* and *irreversibility* of the climate change threat. If the race is lost, humanity will find itself in a radically changed environment. Such metaphors are also applied to construct political arguments: for example, it is argued that particular countries 'lag behind' or that industrialized countries must 'move first.'

In the case of the climate change and security discourse, the end of this horizontal path – of the time landscape – is often represented by the metaphor of a *tipping point* (of +2°C average warming).[23] These collective symbols are particularly important because they help to make sense of the temporality of the climate threat and are needed for the prominent distinction between a linear climate change (horizontal path) and a non-linear chaotic climate change (end of the path). One defining feature of a dangerous climate change is precisely its non-linearity, that is, a development, which is not gradual but erratic and *unpredictable*.

A second concept of movement constructs climate change and related risks in the vertical dimension: as a concrete substance that is rising: "the danger is mounting" (Beck in UNSC 2011b: 27); "we were on the cusp of a global climate disaster without an urgent and ambitious collective response" (Goddard in UNSC 2011b: 29). This concept draws on the metaphors that *the earth is a container* (for states, people, nature etc.) and *above is more; below is less*. One expression of this concept, which is *not* metaphorical is that of sea-level rise. The concept of a vertical movement is crucial for linking the development of different dangers: growing heat-level → rising sea levels → increasing risks (e.g. of getting ill, becoming victim to a natural disaster). The General Assembly debate on climate security in 2009 illustrates these linkages very clearly. The fate of the small island states was treated as a pars pro toto for the fate of humanity in the face of climate change in general: "We will all suffer sooner or later. No country is immune to the impacts of climate change, and the smallest of the SIDS will be the first" (Moses in UNGA 2009b: 2). Other than the vertical concept, however, there is no critical threshold in this type of movement; it simply depicts a linear increase. In conclusion, the vertical axis deals with quantitative changes while the horizontal concept is more concerned with the quality of change. Together these two topical concepts of movement create a specific notion of urgency and provide the discursive background for the precautionary principle that should guide climate politics (UNCED 1992: Art.3).

A further set of important collective symbols is made up of biological metaphors: *states are bodies*; *the earth is a body, climate change is a disease.* "Megacities are a breeding ground for social problems" (Bittner 2007); "close to the heart of small island states" (Sethi 2011); "the Earth's life-support systems" (Steiner in UNSC 2011a: 4); "conflict is perhaps the most alarming symptom" (Sanders 2009). There is a strong affinity between the war and health metaphors that stems from their common roots in human experience with the fragile physicality of the body and the precariousness of life (Sarasin 2003: 191–192). The prominence of health metaphors in political discourse is related to the body metaphor in politics, which can be traced back to ancient thought and was popularized in modern political thinking, especially through Hobbes' *Leviathan* (see e.g. Harvey 2007: 6–10; Rigotti 1994).[24]

In international politics, those collective symbols are prominent because they allow not only collectives such as nation states to be treated like individual actors but, moreover, a certain *topology of danger* to be constructed. As such, they identify potential harms, pathogenic agents or vulnerable parts of the earth's organism. Further topological concepts can be found in the construction of risky spaces through metaphors like climate 'hotspots' (Rubner 2011), a 'belt of tension' (Lee 2009) 'flashpoints' (Sanders 2009) or an 'arc of crisis' (Bittner 2007, author's translation).

Finally, different metaphors and symbols revolving around religious or political notions of *apocalypse and damnation* characterize the deep structure of the climate security discourse.[25] Such metaphors link up with the motional metaphors presented above and construct the manifestation of dangerous climate change as an endpoint or limit to the world as we know it. Recalling apocalyptic fears from nuclear holocaust during the Cold War, such collective symbols demonstrate the magnitude of the climate threat, resulting in "nations battered by the fury of a deadly hurricane of biblical proportions" (Flores in UNSC 2011b: 12) or even "the exodus of populations" (Ragaglini in UNSC 2011b: 41), but also argue for "the historic and overriding need to take decisive action to assuage the scourge" (Rodríguez in UNSC 2011b: 23). Religious collective symbols link up with the motion metaphors described above, thereby constructing a *teleology* of climate change: "we marshaled the internal strength we needed to overcome this harsh test and move on, but not without having to retrace our steps along the same path" (Flores in UNSC 2011b: 12). These collective symbols that foretell an apocalyptic future further increase the antagonistic framing of war metaphors by presenting climate change as a total threat to the whole of humanity (Methmann and Rothe 2012: 327–330).

From mechanics to complexity science

Another category of collective symbols has its roots in physics and other natural sciences and is applied to express causalities and to make sense of international politics (see also Jäger and Jäger 2007: 56–57). Nation states in such a concept are *containers* that are able to hold together the forces they contain: e.g. a population lives *within* a nation (Charteris-Black 2006: 575). In post-Cold War security discourse, the *container* metaphor became one reason for the problematization of weak or fragile states. If the walls of the container become *porous, weak* or *fragile*, dangerous 'elements' of failed states such as pathogenic, criminal or radical subjects could easily 'leak out' or 'spill over' into the broader regional or international space, as exemplified in the following quote: "The effects will spill over borders" (Cape Times 2008).

Further concepts are applied to present complex causalities in simple, almost technical terms. The construction of migration in the discourse is very illustrative in this respect. For example, migration is often expressed in terms of 'push and pull factors'; 'streams of migration'; and being 'triggered' or 'amplified.'[26] Such expressions can also be found throughout the sample to depict other threats from climate change. In the following quote, physical metaphors are linked to the notion of the state as a container: "What is certain is that migration is going to increase, and if you close the borders to Africa, you will even amplify the situation there. This is like a *pressure cooker*" (Bittner 2007, author's translation). Such physical or technical expressions allow very simple and direct correlations to be drawn between very complex social, political and economic phenomena, as illustrated in this quote: "Such *pressures* could also lead to more disputes over territory and water supplies, exacerbating social and religious *tensions* and *fueling* the radicalization of the poor" (Castle 2008, author's translation and emphasis). Such technical or physical expressions naturalize social phenomena, e.g. by blurring the subject status of actors such as the poor or the migrants, and by constructing over-simplistic relatively linear causalities between complex issues.

Further prominent collective symbols in climate change and security discourse are taken from more recent theoretical developments in mathematics and the natural sciences, such as complexity theory (see Saperstein 1997): 45–46). Metaphors stemming from these discourses depict chaos or turbulence as illustrated in this example: "The chaos that results can be an incubator of civil strife" (Knickerbocker 2007). Again motional metaphors play a big role, as all of the perceived complex threats are marked by a particularly rapid motion, such as in this example: "Sudden shocks can tip fragile countries into conflict and instability" (DfID *et al.* 2011: 7).

The container or vehicle model of the state – old and new

Western liberal political discourses have developed a very particular discursive vocabulary around a few important metaphorical clusters: body/cleanliness/infection/war and struggle/invasion/otherness or migration (Sarasin 2003: 227). These symbols are deeply sedimented within our political culture and can be invoked in concrete discursive struggles. These metaphors and symbols are extremely important in processes of securitization as they are used to construct antagonisms: the inner community for example is constructed as a body; the political enemy for example as an invasion of pathogens.[27] In the 1980s a group of critical discourse scholars around German linguist Werner Link have collected such collective symbols that make up Western political discourses through a whole series of empirical investigations. Out of this collection they developed a model that illustrates how these collective symbols are linked to constitute a broader deep structure or political discourse (Jäger and Jäger 2007: 41–42). I have slightly revised this *container, body or boat model* of the political community to adapt it to my case. It helps in making sense of the deep structure of Western security discourses and is summarized in Figure 4.3.

What one can see from the figure is first that two topical axes help make sense of the political space and political developments. The horizontal axis helps for example make sense of time as a linear process/way (e.g. 'dangers that lie in front of us'). A vertical axis might express value judgments or judgments about developments (e.g. 'as unemployment rises, so

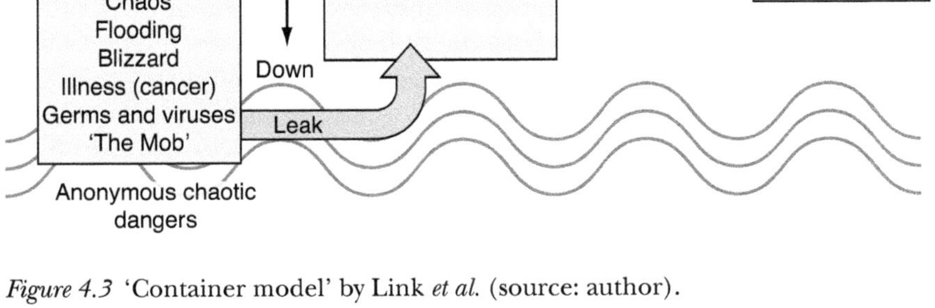

Figure 4.3 'Container model' by Link *et al.* (source: author).

does criminality'). Second, political communities metaphorically are constructed as containers, bodies, or vehicles that can move forwards (and backwards). And they have a clear-cut boundary to the outside, a 'mantle' or 'hull.'

What is important to the study of securitizations is that these vehicles or bodies are metaphorically located in a dangerous vicinity – symbolized in the graphic by the rough sea. Within this dangerous environment there are two types of dangers: on the one hand, there are enemies, i.e. counter-communities with a subject status that threaten the inside community. This construction of antagonism equals a security discourse along the lines of a traditional security concept. On the other hand, the surrounding sea hosts a whole range of subject-less dangers: from political chaos and 'the mob,' to diseases, cyber-threats or disasters. There is a common pool of collective symbols to represent these potential risks for a political community. For example economic risks, just like international migration or risks of a cyber attack, are often constructed in terms of natural disasters. This means that there is already a certain discursive affinity between those different risks, which makes it easier to link them argumentatively. As the vehicle can never be completely isolated from such dangers, there is always the danger that they might leak into the political container of the community. Such a leak might manifest itself as inner chaos, political subversion, spread of crime or as an intrusion of migrants. What the container model by Link *et al.* shows is that conceptual metaphors and symbols in security discourse are closely related and constitute broader symbolic structures, which set the limit of the sayable and thinkable in security discourses.

I have compared the original container model developed by Link *et al.* in the 1980s with the current discourse on climate security in order to identify differences in the symbolic structure and peculiarities of the climate security discourse. Based on the collective symbols identified in climate security discourse I have developed a revised version of the container or vehicle model of the state, which is presented in Figure 4.4.

In climate change and security discourse political communities are still represented as containers or vehicles. Two different sorts of dangers threaten the political body/container. On the one hand, metaphors of personification construct climate change as an enemy of humanity (with a subject status). On the other hand, there is a range of subject-less, anonymous and blurry risks such as diseases, political chaos, terrorism and migration. The container or vehicle of a political community is situated in this dangerous environment – often symbolized through the metaphor of a rough sea. Motional metaphors along vertical and horizontal axes indicate changes in this dangerous environment. Vertically, the dangers surrounding the state are rising: sea level, instability, the risk of extreme weather, etc. On a horizontal axis, the states as vehicles are in a race against dangerous climate change. If climate change is moving too fast,

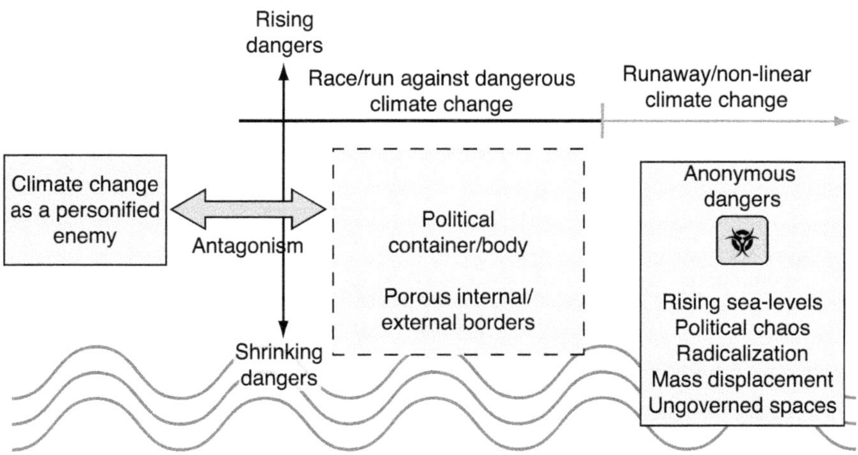

Figure 4.4 Revised container model: collective symbols in climate security discourse (source: author).

the earth and national boats are going to find themselves in troubled waters. Dangerous climate change is acting as a game changer, then, making the sea of surrounding dangers unpredictable to navigators. This becomes obvious from the following quote:

> We do not know precisely where particular tipping points lie; we know they exist. Like a ship navigating through the fog we need to make a judgment about how close we go towards the rocks in order to shorten the route to our destination.
>
> (Mabey 2011: 43)

At the same time, political reality (at least perceived) has become so *complex* that the outside walls of the container have become porous or damaged. This means that more and more dangerous elements can leak into the political community. In the worst case scenario one cannot distinguish between inside and outside (or non-metaphorical internal and external security) anymore. What the container or body model shows is that the collective symbols of climate change and security discourse are consistently linked to established political metaphors.[28] For example "the symbiotic relationship between biomedical discourses and strategic discourses" (Campbell 1992: 80) exemplifies the close ties that exist between different conceptual fields. The discursive deep structure is the discursive instance that links such handed-down concepts to modern political collective symbols, together allowing coherent narratives to be constructed.

The interlinked metaphorical deep structure of the climate change and security discourse is summarized in Figure 4.5. In concrete discursive

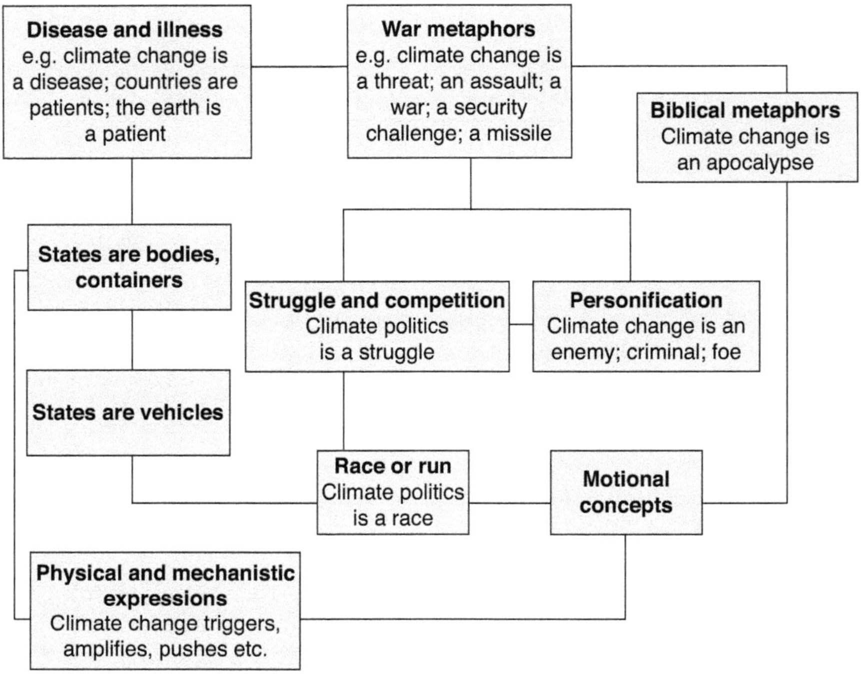

Figure 4.5 Metaphorical deep structure of the climate change and security discourse (source: author).

struggles, speakers intentionally or unconsciously draw on this metaphorical deep structure. Collective symbols are used to make sense of climate change and security issues, represent causal relations, or ethical statements. The metaphor of *climate change as a race*, for example, implies a call for urgency of policy responses; physical and other expressions from natural sciences construct simple, mechanistic causalities; the metaphor of climate change as an enemy helps to make sense of the complex phenomenon of global warming. The existence of war metaphors in the climate change and security discourse does not yet mean a securitization of the issue. Nevertheless, if two problems are constructed through the same collective symbols (if climate change is problematized in the same terms as other security risks such as for example terrorism), it becomes easier to link them argumentatively. In this respect, war metaphors represent a *securitization precursor*.

Furthermore, as many of the collective symbols belong to the inventory of interdiscourse, they produce discursive affinities between climate change and other non-traditional security threats. Through analogies and metaphorical transfers among dangers such as crime, war, natural catastrophes, diseases and pandemics, this deep structure produces an epistemological disposition, drawing on discursive affinities, that influences how we perceive these problems. In sum, the interrelated structure of collective symbols outlined in this chapter allows climate change and security storylines to be narrated in a coherent and meaningful way. In the coming three chapters we will see how such symbols are applied in discursive struggles and condensed into broader storylines or narratives.

Notes

1 Further research groups on environment and security developed – partly through a critique of the Toronto group – with the ENCOP project in Zurich (Spillmann and Bächler 1995) and the Global Environmental Change and Security Project (GECHS), an international research project founded in 1996, which was at first based at the University of California in Irvine, later at the University of Oslo (Lonergan 1999).

2 At the center of this debate was the question whether civil conflicts in developing countries are driven by atypical economic opportunities generating greed among societal/ethnic groups or by atypical suffering, i.e. social and economic inequalities, leading to grievances of marginalized social groups. The greed-vs.-grievance debate is related to the one on environmental conflict, since 'greed,' in this line of argument, is driven by the abundance of resources, whereas grievances can be fuelled by the degradation of resources worsening the living conditions of 'the poor.'

3 For example, in 2003, NATO organized a high-level workshop entitled *Desertification in the Mediterranean region: A security issue?* in Valencia, Spain (Kepner and Rubio 2003). In December 2007, NATO in cooperation with the OSCE held a further workshop on *Water Scarcity, Land Degradation and Desertification in the Mediterranean region – Environment and security aspects* (see Kalbermatten 2009).

4 To be precise here: not every desert is an ungoverned space – there are major cities and different forms of political organization in many desert areas. What is problematized is missing forms of statehood in a Western sense.

5 Bureaucrats and diplomats from different Spanish ministries have been particularly active in promoting a securitization of desertification, for example in the UN Convention to Combat Desertification (Brauch and Spring 2009: 11).

6 While desertification and climate change are, of course, physically linked, their concrete articulation is contingent: in one case climate change might be constructed as the major concern while desertification is subsumed as one particular manifestation of it; in another case desertification might be framed as the major threat while climate change is articulated as one of its causes.

7 A database search in German newspapers, based on LexisNexis, using the keywords 'climate change' (Klimawandel) AND 'security' (Sicherheit) produced 35 hits for the whole of the 1990s, while none of the articles actually referred to the issue of climate change and security in a narrow sense. The same search applied for the year 2007 alone produced 1177 hits, among them many articles directly dealing with the security implications of climate change.

8 As Oreskes famously noted, as early as the first IPCC report in 1988, a vast consent on anthropogenic climate change already marked the scientific discourse. Political and media discourses instead constructed a picture of a highly contested scientific debate (Oreskes 2004).

9 Al Gore as vice president to Bill Clinton in the 1990s had been strongly influenced by the arguments of Robert Kaplan's *The Coming Anarchy* (Phone interview with a climate security researcher, September 29, 2011). Kaplan (1994)

projected how problems such as resource scarcity, crime or disease would threaten the "social fabric of the planet."

10 The section on Africa, for example, comprised 11 references to violent conflict – a novelty for the IPCC that was otherwise very cautious about this issue (Gleditsch and Nordås 2009: 21).

11 Besides the already mentioned USA, UK and Germany, these states were Sweden, Denmark, the Benelux countries, Canada and Australia (see e.g. Interview with a climate security analyst, Berlin September 14, 2011; Phone interview with a Geneva-based climate security analyst, September 13, 2011).

12 An exception are statements by UNEP head Achim Steiner, who, for example, gave a speech in the last Security Council debate on climate change (Steiner in UNSC 2011a: 3–6).

13 A prominent example of such regional activities is the Environmental Security Initiative run by UNDP, UNEP, NATO, OSCE, and the UN Economic Commission for Europe which deals with climate change as a focus area and should "encourage key public decision-makers in South Eastern and Eastern Europe, Central Asia and the Caucasus to act to advance and protect peace and the environment at the same time" (UNEP 2012).This is done through joint workshops, the transfer of knowledge, assessment methods and expertise.

14 This is due to the fact that some of its members are less interested in environmental concerns and do not consider them as NATO tasks. NATO military staff thus cannot officially address the problem of climate change (Phone interview with a Brussels-based security analyst, October 25, 2011). Nevertheless, divisions such as the Science for Peace Program or the Emerging Security Challenges Division are allowed to work conceptually on the topic.

15 For example, the emergency-planning unit in August 2011 held for the first time a meeting on extreme weather events (Phone interview with a Brussels-based security analyst, October 25, 2011).

16 The IPCC distinguished between North Africa, southern Europe and West Asia, all of which displayed vulnerabilities against particular negative climate change impacts (Watson *et al.* 1997: 8–10). The Working Group II report on *Impacts, Adaptation and Vulnerability* on the contrary did not adopt a regional perspective but assessed climate change risks in different continents (IPCC 2007).

17 In 2010 the World Bank, for example, organized the first *Workshop on Climate Induced Migration and Displacement in the Middle East and North Africa* (Worldbank 2010).

18 The ESS only acknowledges a general link between climate change and a growing competition for national resources (EU Security Strategy 2003: 2–3).

19 For example, in the UNGA and UNSC debates on climate change and security, MENA countries' representatives supported the position of the G77, which resisted understanding climate change in terms of security and instead understood it as a development problem.

20 Also instructive is Eskjaers' observation that media coverage explicitly deals with climate change as an issue of international politics, but there are almost no regional perspectives in news reports on climate change (Eskjaer 2009: 361).

21 The phenomenon of climate change cannot be perceived in an immediate sense. Although the phenomenon is real and cannot reasonably be denied any longer today, it is nevertheless a scientific construction based on a whole variety of numbers, models, calculations, visibilities. Climate is a pattern, an arithmetic mean or a trend; this is something entirely different from the actual weather, which can be felt by humans in a direct manner.

22 As for example Koller demonstrates, concepts of sport – and other competitions – are socio-culturally closely related to concepts of war (Koller 2004: 13).

23 A similar popular metaphor that illustrates this spatial framing of climate change is the metaphor of a carbon lock-in, as demonstrated here: "In the short term, we are locked into global warming" (Howden 2008).

24 The most radical expression of a body/health metaphor is found in the genocidal, anti-Semitic ideology of German Fascism. The Nazis had conceptualized the German state as a collective body that had to be cured of the Jewish parasites (see Musolff 2010).

25 As John Gray argues, apocalyptic thinking is not only a crucial feature of Christianity (and other major religions) but can be found in all major political ideologies in modernity (Gray 2007). While the triumph of liberalism after the Cold War seemingly marked the end of apocalyptic political ideology, as Fukuyama's prognosis of 'the end of history' (Fukuyama 1989) suggests, discourses such as that of climate change indicate a renaissance of apocalyptic thinking.

26 One certainly has to acknowledge that these expressions are strongly conventionalized, i.e. other than, for example, war metaphors, their figurative meaning is not recognized anymore. Nevertheless, their conventional meaning is strongly influenced by natural sciences and a mechanistic causality. The term *tension* is a good example: it is commonly used to denote social phenomena, i.e. a situation in which a conflict might arise. Originally, however, it refers to a stretched condition of a material, later also to an electromotive force (Online Etymology Dictionary 2015). Although the use of tension in the social realm is highly conventional today, these earlier meanings are still present in its current use.

27 One can find these collective symbols in different discursive fields such as migration and terrorism, and even in the climate change discourse as I will show in this study.

28 The ship or boat metaphor as a symbol for the state has its roots in ancient (Egyptian and Greek) political thinking, just as the rough sea is a symbol of threats to the community or the statesmen as navigators (Schäfer 1972).

References

9–11 Commission. 2004. "The 9/11 commission report: Final report of the national commission on terrorist attacks upon the United States." US National Commission on Terrorist Attacks Upon the United States. Accessed May 4, 2015. www.9-11commission.gov/report/911Report.pdf.

Abrahamsen, Rita. 2004. "A breeding ground for terrorists? Africa and Britain's 'war on terrorism'." *Review of African Political Economy* 31 (102): 677–684.

Abrahamsen, Rita. 2005. "Blair's Africa: the politics of securitization and fear." *Alternatives: Global, Local, Political* 30 (1): 55–80.

Adeel, Zafar, Janos Bogardi, Christopher Braeuel, Pamela Chasek, Maryam Niamir-Fuller, Donald Gabriels, Caroline King, Friederike Knabe, Ahang Kowsar, Boshra Salem, Thomas Schaaf, Gemma Shepherd and Richard Thomas. 2006. *Overcoming one of the greatest environmental challenges of our times: Re-thinking policies to cope with desertification.* Bonn: United Nations University.

Adger, Neil W., Tor A. Benjaminsen, Katrina Brown and Hanne Svarstad. 2001. "Advancing a political ecology of global environmental discourses." *Development and Change* 32 (4): 681–715.

Barnett, John. 2001. *The meaning of environmental security: Ecological politics and policy in the new security era.* New York: Zed Books.

Bittner, Jochen. 2007. "Die Klima-Kriege." *Die Zeit,* May 3, 2007. Accessed April 30, 2015. www.zeit.de/2007/19/Klimawandel.

Boas, Ingrid. 2015. *Climate migration and security. Securitisation as a strategy in climate change politics.* New York: Routledge.

Brauch, Hans G. 2002. "Climate change, environmental stress and conflict." In *Climate change and conflict,* edited by the Federal Ministry for the Environment, 9–112. Berlin.

Brauch, Hans G. 2012. "Policy responses to climate change in the Mediterranean and MENA Region during the Anthropocene." In *Climate change, human security and violent conflict,* edited by Jürgen Scheffran, Michael Brzoska, Hans Günter Brauch, Peter M. Link and Janpeter Schilling, 719–794. Berlin, Heidelberg: Springer.

Brauch, Hans G. and Úrsula O. Spring. 2009. "Securitizing the ground, grounding security." In *Desertification land degradation and drought: UNCCD issue paper No. 2.* United Nations Conference to Combat Desertification.

Brauch, Hans G., Antonio Marquina, Mohammad El Sayed-Selim, Peter H. Liotta and Paul F. Rogers. 2003. *Security and environment in the Mediterranean: conceptualising security and environmental conflicts.* Berlin: Springer.

Brown, Oli. 2007. "Weather of mass destruction? The rise of climate change as the 'new' security issue." Winnipeg: International Institute for Sustainable Development (IISS).

Brundtland Commission, The World Commission on Environment and Development. 1987. *Our common future.* Oxford: Oxford University Press.

Cabinet Office. 2008. National security strategy of the United Kingdom: Security in an interdependent world. Crown Copyright Cm 7291. Norwich: Cabinet Office.

Callick, Rowan. 2011. "UN snubs pleas on climate change." *The Australian,* July 22, 2011, 8.

Campbell, David. 1992. *Writing security. US foreign policy and the politics of identity.* Minneapolis: University of Minnesota Press.

Campbell, Kurt M., Jay Gulledge, J. R. McNeill, John Podesta, Peter Ogden, Leon Fuerth, R. J. Woolsey, Alexander T. Lennon, Julianne Smith and Richard Weitz. 2007. *The age of consequences: the foreign policy and national security implications of global climate change.* Washington DC: Center for Strategic and International Studies.

Cape Times. 2008. "Think about climate change as a major threat to security, not old-style warfare, researcher says." October 20, 2008, 3.

Carius, Alexander, Sebastian Oberthür and Dennis Tänzler. 2002. "Climate change, environmental stress and conflict." In *Climate change and conflict,* edited by the Federal Ministry for the Environment, 113–142. Berlin.

Carvalho, Anabela and Jacquelin Burgess. 2005. "Cultural circuits of climate change in UK broadsheet newspapers, 1985–2003." *Risk analysis* 25 (6): 1457–1469.

Castle, Stephen. 2008. "Migration to surge, EU leaders are told." *International Herald Tribune,* March 8, 2008, 3.

Center for Naval Analysis. 2007. "National security and the threat of climate change." Last Modified 2007. Accessed May 4, 2015. https://www.cna.org/sites/default/files/National Security and the Threat of Climate Change – Print.pdf.

Chandler, David. 2007. "The security–development nexus and the rise of 'anti-foreign policy'." *Journal of International Relations and Development,* 10(4): 362–386.

Charteris-Black, Jonathan. 2006. "Britain as a container: immigration metaphors in the 2005 election campaign." *Discourse and Society* 17 (5): 563–581.

Collier, Paul and Anke Hoeffler. 2005. "Resource rents, governance, and conflict." *Journal of conflict resolution* 49 (4): 625–633.

Council, The Council of the European Union. 2009. *Joint progress report and follow-up recommendations on climate change and international security (CCIS) to the Council.* Brussels: European Union.

Dalby, Simon. 2002. *Environmental security.* Minneapolis: University of Minnesota Press.

De Goede, Marieke. 2008. "Beyond risk: Premediation and the post-9/11 security imagination." *Security Dialogue* 39 (2–3): 155–176.

De Soysa, Indra. 2002. "Ecoviolence: shrinking pie, or honey pot?" *Global Environmental Politics* 2 (4): 1–34.

Detraz, Nicole and Michele M. Betsill. 2009. "Climate change and environmental security: For whom the discourse shifts." *International Studies Perspectives* 10 (3): 303–320.

DfID, Department for International Development, Foreign and Commonwealth Office FCO, and Ministry of Defence MoD. 2011. *Building stability overseas strategy.* London: DfID.

Duffield, Mark and Nicholas Waddell. 2006. "Securing humans in a dangerous world." *International Politics* 43 (1): 1–23.

El-Sayed Selim, Mohammad. 2011. "Arab perceptions of soft security issues." In *Coping with global environmental change, disasters and security: Threats, challenges, vulnerabilities and risks,* edited by Hans G. Brauch, Ursula O. Spring, Czeslaw Mesjasz, John Grin, Patricia Kameri-Mbote, Béchir Chourou, Pál Dunay and Jörn Birkmann, 327–338. Berlin: Springer.

Elasha, Balgis Osman. 2010. "Mapping of climate change threats and human development impacts in the Arab region." In *UNDP Arab development report–research paper series.* New York: UNDP Regional Bureau for the Arab States.

Eskjaer, Mikkel F. 2009. "Communicating climate change in regional news media." *International Journal of Climate Change Strategies and Management* 1 (4): 356–367.

EU Security Strategy. 2003. *A secure Europe in a better world: The European security strategy,* drafted by EU High Representative Javier Solana. Brussels: European Council.

Floyd, Rita. 2008. "The environmental security debate and its significance for climate change." *The International Spectator* 43 (3): 51–65.

Fukuyama, Francis. 1989. "The end of history?" In *Globalization and the challenges of a new century: A reader,* edited by Patrick O'Meara, Howard D. Mehlinger and Matthew Krain, 161–180. Bloomington, IN: Indiana University Press.

Georgas, Dimitri. 2003. "Impacts of climate changes and sea level rise on the Mediterranean coastal zone." In *Security and environment in the Mediterranean,* edited by Hans G. Brauch, Peter H. Liotta, Antonio Marquina, Paul F. Rogers and Mohammad El-Sayed Selim, 631–642. Heidelberg: Springer.

Gleditsch, Nils P. 1998. "Armed conflict and the environment: A critique of the literature." *Journal of Peace Research* 35 (3): 381–400.

Gleditsch, Nils P. and Ragnhild Nordås. 2009. "IPCC and the climate-conflict nexus." *IOP conference series: Earth and environmental science* 6.

Gleick, Peter H. 1989. "The implications of global climatic changes for international security." *Climatic Change* 15 (1–2): 309–325.

Gleick, Peter H. 1990. "Climatic changes and international security." *ECO Newsletter* 75 (5): 5.

Government Office for Science. 2011a. *Foresight: migration and global environmental change. Future challenges and opportunities. Final report.* London: Crown copyright. URN 11/1116.

Government Office for Science. 2011b. *Foresight: the international dimensions of climate change. Final project report.* London: Crown copyright. URN 11/1042.

Gray, John. 2007. *Black mass: Apocalyptic religion and the death of utopia.* New York: Farrar, Straus and Giroux.

Hajer, Maarten A. 1995. *The politics of environmental discourse: ecological modernization and the policy process.* Oxford: Clarendon Press.

Hartmann, Betsy. 2006. "Liberal ends, illiberal means national security, 'environmental conflict' and the making of the Cairo consensus." *Indian Journal of Gender Studies* 13 (2): 195–227.

Hartmann, Betsy. 2010. "Rethinking climate refugees and climate conflict: Rhetoric, reality and the politics of policy discourse." *Journal of International Development* 22 (2): 233–246.

Harvey, Arnold D. 2007. *Body politic: political metaphor and political violence.* Newcastle: Cambridge Scholars Publishers.

Helman, Gerald B. and Steven R. Ratner. 1992. "Saving failed states." *Foreign policy* (89): 3–20.

Homer-Dixon, Thomas F. 1991. "On the threshold: environmental changes as causes of acute conflict." *International Security* 16 (2): 76–116.

Homer-Dixon, Thomas F. 1994. "Environmental scarcities and violent conflict: evidence from cases." *International Security* 19 (1): 5–40.

Homer-Dixon, Thomas F. 1999. *Environment, scarcity, and violence.* Princeton NJ: Princeton University Press.

Houghton, John. 2003. "Global warming is now a weapon of mass destruction." *The Guardian,* July 28, 2003.

Howden, Daniel 2008. "Climate change 'hits minorities hardest'." *The Independent,* March 11, 2008, 34.

IOM, International Organization for Migration. 2009. "Compendium of IOM's activities in migration, climate change and the environment." Accessed April 30, 2015. https://www.iom.int/jahia/webdav/shared/shared/mainsite/activities/env_degradation/compendium_climate_change.pdf.

IPCC. 2007. "Summary for policymakers." In *Climate change 2007: impacts, adaptation and vulnerability. Contribution of working group II to the fourth assessment report of the intergovernmental panel on climate change,* edited by M. L. Parry, O. F. Canziani, J. P. Palutikof, P. J. van der Linden and C. E. Hanson, 7–22. Cambridge: CUP.

Jackson, Robert H. 2000. *The global covenant: human conduct in a world of states.* Oxford: OUP.

Jäger, Margarete and Siegfried Jäger. 2007. *Deutungskämpfe, Theorie und Praxis Kritischer Diskursanalyse.* Wiesbaden: VS Verlag für Sozialwissenschaften.

Kalbermatten, Grégoire. 2009. "Grounding security, securing the ground." In *Water scarcity, land degradation and desertification in the Mediterranean region,* edited by José L. Rubio, Uriel Safriel, Raul Daussa, Winfried Blum and Fausto Pedrazzini, 19–32. Berlin: Springer.

Kaldor, Mary, Mary Martin and Sabine Selchow. 2007. "Human security: a new strategic narrative for Europe." *International Affairs* 83 (2): 273–288.

Kaplan, Robert. 1994. "The coming anarchy: how scarcity, crime, overpopulation, tribalism, and disease are rapidly destroying the social fabric of our planet." *The Atlantic Magazine.*

Kepner, William G. and Jose L. Rubio. 2003. Desertification in the Mediterranean region: a security issue. NATO – CCMS and science committee workshop, Valencia, Spain December 2–5, 2003. Executive summary. Brussels: North Atlantic Treaty Organization (NATO).

Ki-moon, Ban. 2007. "A climate culprit in Darfur." *The Washington Post*, April 16, 2007. Accessed April 30, 2015. www.washingtonpost.com/wp-dyn/content/article/2007/06/15/AR2007061501857.html.

Klare, Michael T. 2007. "Global warming battlefields: how climate change threatens security." *Current History* 106 (703): 355–361.

Knickerbocker, Brad. 2007. "Could warming cause war?" *Christian Science Monitor*, April 19, 2007, 2.

Koller, Veronika. 2004. "Businesswomen and war metaphors: 'possessive, jealous and pugnacious'?" *Journal of Sociolinguistics* 8 (1): 3–22.

Kühn, Florian P. 2010. *Sicherheit und Entwicklung in der Weltgesellschaft, Liberales Paradigma und Statebuilding in Afghanistan.* Wiesbaden: VS Verlag.

Lakoff, George and Mark Johnson. 1980. *Metaphors we live by.* Chicago: University of Chicago Press.

Lakoff, George, Jane Espenson and Alan Schwart. 1991. *Master metaphor list.* Berkeley CA: University of California.

Lee, James R. 2009. "Global warming is just the tip of the iceberg." *Washington Post*, January 4, 2009, 3.

Lonergan, Steve. 1999. *Global environmental change and human security: GECHS science plan.* Bonn: International Human Dimensions Programme on Global Environmental Change.

Lutes, Mark W. 1998. "Global climate change." In *Political ecology: global and local*, edited by Rodger Keil, David V. J. Bell, Peter Penz and Leesa Fawcett, 157–175. London: Routledge.

Maas, Achim and Dennis Tänzler. 2009. *Regional security implications of climate change. A synopsis*. Berlin: Adelphi research.

Mabey, Nick. 2008. *Delivering climate security: International security responses to a climate changed world. Edited summary.* London: E3G.

Mabey, Nick. 2011. *Degrees of risk: defining a risk management framework for climate security.* London: E3G.

MacFarquhar, Neil. 2009. "Islanders fearing climate change press a UN debate." *New York Times*, May 29, 2009, 4.

Matthew, Richard A. 1995. "Environmental security: demystifying the concept, clarifying the stakes." *Woodrow Wilson Environmental Change and Security Project Report*: 14–23.

Mathews, Jessica Tuchman. 1989. "Redefining security." *Foreign affairs*: 162–177.

Meadows, Dennis L., Donella H. Meadows, Jorgen Randers and William W. Behrens. 1972. *The limits to growth.* New York: Universe Books.

Meeks, Gregory and Michael Shank. 2008. "Security Council must act preemptively – on climate change." *Christian Science Monitor*, March 24, 2008, 9.

Methmann, Chris and Angela Oels. 2015. "From 'fearing' to 'empowering' climate refugees: Governing climate-induced migration in the name of resilience." *Security Dialogue* 46 (1): 51–68.

Methmann, Chris P. and Delf Rothe. 2012. "Politics for the day after tomorrow:

the political effect of apocalyptic imaginaries in global climate governance." *Security Dialogue* 43 (3): 323–344.

Mitchell, Katharyne. 2010. "Ungoverned space: global security and the geopolitics of broken windows." *Political Geography* 29 (5): 289–297.

Musolff, Andreas. 2010. *Metaphor, nation and the holocaust: the concept of the body politic.* New York: Routledge.

NIC, National Intelligence Council. 2009. *North Africa: the impact of climate change to 2030: geopolitical implications.* Washington, DC: NIC.

Online Etymology Dictionary. 2015. "Tension." Accessed April 30, 2015. www.etymonline.com/index.php?term=tension&allowed_in_frame=0.

Oreskes, Naomi. 2004. "The scientific consensus on climate change." *Science* 306 (5702): 1686–1686.

Piazza, James A. 2006. "Rooted in poverty? Terrorism, poor economic development, and social cleavages." *Terrorism and Political Violence* 18 (1): 159–177.

Rigotti, Francesca. 1994. *Die Macht und ihre Metaphern: über die sprachlichen Bilder der Politik.* Frankfurt/Main: Campus-Verlag.

Rogers-Hayden, T., F. Hatton and I. Lorenzoni. 2011. "'Energy security' and 'climate change': Constructing UK energy discursive realities." *Global Environmental Change* 21 (1): 134–142.

Rotberg, Robert I. 2002. "Failed states in a world of terror." *Foreign Affairs* 81: 127–140.

Rubner, Jeanne. 2011. "UN-Studie: Erderwärmung verschärft Konflikte in Afrika." *Süddeutsche Zeitung,* November 26, 2011, 4.

Sanders, Edmund. 2009. "Shrinking resources ignite the African wars of climate change." *The Independent South Africa,* December 6, 2009, 17.

Saperstein, Alvin M. 1997. "Complexity, chaos, and national security policy: metaphors or tools." In *Complexity, global politics, and national security,* edited by David Alberts and Thomas J. Czerwinski, 44–61. Washington, DC: National Defense University.

Sarasin, Philipp. 2003. *Geschichtswissenschaft und Diskursanalyse.* Frankfurt am Main: Suhrkamp.

Schäfer, Eckart. 1972. "Das Staatsschiff: Zur Präzision eines Topos." In *Toposforschung,* edited by Peter Jehn, 259–292. Frankfurt/Main: Athenäum.

Schwartz, Daniel M. 2003. "Environmental security post September 11th." *Journal of Environmental Peace* 2.

Schwartz, Peter and Doug Randall. 2003. *An abrupt climate change scenario and its implications for United States national security.* Washington, DC: Global Business Network.

Sethi, Nitin. 2011. "India prevents UNSC from hijacking climate talks." *The Times of India,* July 23, 2011. Accessed April 30, 2015. http://timesofindia.indiatimes.com/home/environment/developmental-issues/India-prevents-UNSC-from-hijacking-climate-talks/articleshow/9331057.cms.

Smith, Dan and Janani Vivekananda. 2007. *A climate of conflict: the links between climate change, peace and war.* London: International Alert.

Smith, Paul J. 2007. "Climate change, mass migration and the military response." *Orbis* 51 (3): 617–633.

Solana, Javier. 2008. *Climate change and international security.* Paper from the High Representative and the European Commission to the European Council. Brussels: European Commission.

Spiegel Online. 2009. "Klimawandel: Forscher warnen vor beispiellosen Flüchtlingswellen." Last Modified June 10, 2009 Accessed April 30, 2015. www.spiegel.de/wissenschaft/natur/klimawandel-forscher-warnen-vor-beispiellosen-fluechtlings wellen-a-629680.html.

Spillmann, Kurt R. and Günther Bächler. 1995. "Environmental crisis: regional conflicts and ways of cooperation." *Occasional Papers.* Zurich: Environment and Conflicts Project (ENCOP).

Stern, Nicholas. 2006. *The economics of climate change: the Stern review.* Cambridge: CUP.

Stripple, Johannes. 2005. *Climate change after the international: rethinking security, territory and authority.* Lund, Sweden: Department of Political Science, Lund University.

Tänzler, Dennis, Achim Maas and Alexander Carius. 2010. "Climate change adaptation and peace." *Wiley Interdisciplinary Reviews: Climate Change* 1 (5): 741–750.

Trombetta, Maria J. 2012. "Climate change and the environmental conflict discourse." In *Climate change, human security and violent conflict,* edited by Jürgen Scheffran, Michael Brzoska, Hans G. Brauch, Peter M. Link and Janpeter Schilling, 151–164. Heidelberg: Springer.

UNCED, United Nations Conference on Environment and Development. 1992. "United Nations framework convention on climate change." Accessed April 30, 2015. http://unfccc.int/resource/docs/convkp/conveng.pdf.

UNDP. 1994. *Human development report: new dimensions of human security.* New York: United Nations.

UNEP. 2012. "Linking environment and security." Accessed May 5, 2015. www.unep.org/roe/KeyActivities/EnvironmentalSecurity/tabid/54360/Default.aspx.

UNGA, United Nations General Assembly. 2009a. *Climate change and its possible security implications: Report of the Secretary-General.* New York: United Nations General Assembly.

UNGA, United Nations General Assembly. 2009b. Sixty-third session, 85th plenary meeting Wednesday, June 3, 2009.

UNSC, United Nations Security Council. 2007a. Sixty-second year, 5663rd meeting, Tuesday, April 17, 2007.

UNSC, United Nations Security Council. 2007b. Sixty-second year, 5663rd meeting, Tuesday, April 17, 2007, Resumption.

UNSC, United Nations Security Council. 2011a. Sixty-sixth year, 6587th meeting, Wednesday, July 20, 2011.

UNSC, United Nations Security Council. 2011b. Sixty-sixth year, 6587th meeting, Wednesday, July 20, 2011, Resumption.

Watson, Robert T., Marufu C. Zinyowera and Richard H. Moss. 1997. *The regional impacts of climate change: an assessment of vulnerability.* Geneva: Intergovernmental Panel on Climate change (IPCC).

WBGU, German Advisory Council on Global Change. 2009. *Climate change as a security risk.* London: Earthscan.

Westing, Arthur H. 1994. "The Almería statement on desertification and migration: international symposium on desertification and migration, Almería, Spain, 9–11 February 1994." *Environmental Conservation* 21 (2): 179–181.

Worldbank. 2010. "Press release: first workshop on climate induced migration and displacement in the Middle East and North Africa." Accessed April 30, 2015. www.worldbank.org/en/news/press-release/2010/06/15/first-workshop-on-climate-induced-migration-and-displacement-in-the-middle-east-and-north-africa.

5 The international arena

A war of all against nothing

> We must make no mistake. The facts are clear. Climate change is real, and it is accelerating in a dangerous manner. It not only exacerbates threats to international peace and security, it is a threat to international peace and security.
>
> (UN Secretary General Ban-Ki Moon in UNSC 2011a: 2)

At least since the UN Security Council adopted the issue of climate change and energy security in April 2007, the discourse has entered the stage of international politics. But the Security Council did not remain the sole body in global governance that engaged with the discourse. In the previous chapter I outlined in detail how climate change and security was brought to the tables of many different international and regional organizations, among them UN organs such as UNDP and UNEP, security organizations such as NATO and OSCE, and rather specialized actors such as the IOM or the World Bank. While the literature agrees that there has been a massive expansion of the discourse since 2007, the policy implications of this securitization process remain yet unclear. The aim of the chapter is thus twofold: first, I try to explain how and why security could become a dominant frame in international discourses on climate change by tracing the emergence of a climate security discourse coalition (CSDC). Second, I provide an insight into the political effects on the governmental landscape of global (climate) governance.

The emergence of a climate security discourse coalition

Climate change and security discourse developed genealogically out of a broader discourse on environmental conflicts and security as we have seen in the previous chapter. On that note I start my quest for the origins of a CSDC in the field of environmental security. The *nucleus of a transnational CSDC* (Table 5.1) was a smaller number of academic scholars working in the field of environmental security in the early 2000s. They emancipated climate change from broader environmental considerations by discussing

Table 5.1 The nucleus of a transnational CSDC in the early 2000s

Factions	*Subject positions*	*Binding storylines; boundary objects*
Rationalists	Academics Bureaucrats	Redefinition of security Climate change and conflict Climate-induced migration Threat for low-lying island states
Alarmists	Security officials Think tanks	National security implications Abrupt climate change

it as a security term on its own.[1] From environmental security discourse they borrowed three main questions: whether climate change-induced resource scarcity would lead to violent conflict; how global warming would change the understanding of security itself; and last, if climate change would trigger large scale migration (see Barnett 2003). The climate security discourse, however, did not simply mirror environmental security debates but also introduced novel rationales such as the existential threat posed by rising sea levels for low-lying island states (Barnett 2003: 7) and the fear of a non-linear and irreversible climate change, drawing on the tipping point concept (see Schwartz and Randall 2003).

At about the same time, in the years 2002–2004, officials from different ministries in Germany, the UK and the US became interested in the issue when climate change gained ever more prominence as an issue of international politics, for example through the publication of the third assessment report of the IPCC in 2001. Together these subject positions and their articulations made up *the nucleus* of a CSDC – their *common orientation* towards climate change and security linkages constituted the latter as a genuine political problem.

Yet this common orientation towards climate change and security does not mean that the coalition was marked by overall consensus. Table 5.1 indicates that we can broadly distinguish between two *different factions* that each adopted a different take on climate change and security linkages. Academics, on the one hand, have been very cautious in drawing a direct line between climate change and security threats (Barnett 2003: 11). The academic approach of *carefully* testing empirical correlations or discussing the implications of climate change for a theory of security contradicted more alarmist framings in political discourse. Security officials and strategic thinkers, on the other hand, took a different stance to climate security. They tried to assess the implications of climate change for national politics, which required imagining all *possible risks*, for example through the development of potential threat scenarios (e.g. Schwartz and Randall 2003: 7).

As became clear in the previous chapter, from the mid-2000s on more and more actors from different fields became gripped by climate security narratives. At the same time, the early storylines that made up the discourse coalition were reinterpreted and enriched. A critical discourse moment was the UNFCCC Conference of the Parties (COP)13, 2006 in Nairobi, when the then UN Secretary General Kofi Annan called climate change "one of the greatest challenges of our time" and a "threat to global peace and security" (Spiegel Online 2006, author's translation). This articulation was crucial as it brought another important subject position into the CSDC. Elsewhere I have called this form of subjectivity *climate celebrity* (Rothe 2012: 8).[2] Actors that can be assigned to this subject position and have played a major role throughout the empirical material are, for example, Al Gore, IPPC head Rachid Pachauri, UNEP director Achim Steiner, former UNEP head Klaus Töpfer, Sir Nicholas Stern and UN Secretary General Ban Ki-moon. Such actors drew on the storylines of climate security and used them as weapons for argumentation, for example against climate skeptics, in a broader discursive struggle on anthropogenic climate change that has been gaining momentum since early 2007 (see Rothe 2012: 248–210). They thereby turned climate security from a specialized discourse from a few academics and bureaucrats into an *issue of major international interest*.

A starting point for many *national politicians* or even *national leaders* to deal with the issue of climate change and security was the first UNSC Debate on that topic sponsored by the UK government in April 2007. However, these national leaders in the climate security discourse make up anything but one coherent subject position. Rather, one can distinguish between different subjectivities that were adopted in but also shaped or repositioned through debates on climate change and security. The first of these subject positions is that of *forerunners* (taken up mainly by representatives of a few industrialized countries) in a climate change security discourse, for example expressed through the claim to raise international awareness of the security implications of climate change (Howden 2007). In opposition to the forerunners, there is a subject position I would call *climate security skeptics* (not climate skeptics!). This self-identity has been adopted particularly by major emerging powers. The position of climate security skeptics is established through a particular self-attribution, as expressed in this example: "India prevents SC [Security Council] from hijacking climate change" (Sethi 2011). Besides emerging powers, many developing countries also relate to a position of *climate security skeptics*, such as the representative of Cuba in this articulation: "Climate change is an issue that, by definition, should be discussed under the sustainable development cluster" (Benítez Versón in UNGA 2009b: 11).

Yet this does not apply to all members of the G77. This is due to the fact that many developing countries today consider themselves the *victims of climate change*. This subject position is, on the one hand, constituted by the ascriptions of others actors – namely the industrialized states: "it is the poor and the weakest of countries and societies that are most vulnerable to the onslaught" (Oshima in UNSC 2007b: 29). On the other hand, the subject position of the victims is clearly internalized and reproduced by the respective actors themselves: "We are likely to become the victims of a phenomenon to which we have contributed very little" (Aisi in UNSC 2007b: 28). Particularly important to this subject position of the *victims* are representatives from SIDS. Besides Western industrialized countries, SIDS have been the international political actors that have pushed the debate on climate change and security the most (Interview with a climate security analyst, Washington February 26, 2010) – for example, the 2009 UNGA debate on the issue goes back to an initiative by the Pacific SIDS. Furthermore, their unprecedented vulnerability has brought some novel issues – such as the concept of '*statelessness*' – into the discourse (Moses in UNGA 2009b: 1).

Besides these 'black-boxed' national actors, since 2007 further actors from different subject positions have entered the discourse coalition. *International bureaucrats* – representatives of major international organizations – have started to adopt climate security storylines. What is most important is that these stem from a very broad field of different policy backgrounds – besides major security organizations such as NATO or OSCE, there are also organizations from economic (e.g. the World Bank) and development policy (e.g. UNDP). This shows that climate security storylines resonate with the disposition of very different subjects and thus serve as a nodal point between different specialized discourses. In addition, articulations from different *think tanks* – often foreign policy or security oriented – have increased significantly since 2007.

Think tanks that are working for a rather specialized audience, namely political decision makers and bureaucrats, have to be distinguished from *NGOs*, representing the transnational civil society in climate politics (see Newell 2006: 130). My empirical analysis revealed that from 2007 onward one can also find the subject position of NGOs in climate security discourse. These NGOs stems from quite different fields: Greenpeace, for example represents major environmental NGOs (Tenbrock 2009), while other organizations such as Oxfam, Care International or Christian Aid (Spiegel Online 2009) are working more in a humanitarian field. Last, more and more *academic scholars*, even whole research groups, have started researching the security implications of climate change (e.g. Scheffran *et al.* 2012; Theisen *et al.* 2012). Broadly speaking we can notice that the tone in climate security research remains cautious.[3] Empirical studies generally stress the multi-causal nature of any violent conflict (Phone interview with an Oslo-based climate security researcher, September 29, 2011). The same holds true for *research oriented* think tanks such as the International Institute for Sustainable Development or adelphi research (see Chapter 4). A few scholars even studied and warned against the negative implications of a climate security discourse (see Chapter 1).

The faction of the alarmists

Drawing on these analytical observations, the securitization of climate change can be unmasked as a lengthy and enduring process. The *current climate security discourse coalition* developed organically over time out of the nucleus that I summarized in Table 5.1 by winning over a range of further actors from very different subject positions. These started (re)articulating climate change and security storylines. First of all, *climate celebrities*, *fore-runners* (industrialized countries), *victims* (especially SIDS), *NGOs* like Greenpeace or Christian Aid, and *security think tanks* articulated very alarming storylines of climate change and security (Rothe 2012: 248–210). From a hegemony perspective, the inclusion of the subject positions of the victims as well as humanitarian NGOs into a transnational CSDC is particularly important. It is not sufficient that climate security arguments are being articulated from authoritative voices. Rather, hegemony is created by establishing consensus between the different antagonistic parts of society.

All of these actors individually had very different intentions and reasons to draw on alarmist storylines and present climate change as a security threat: to raise attention for the problem of climate change; to put pressure on industrialized countries; to problematize humanitarian issues; or to anticipate and prevent potential crisis situations. Nevertheless, their articulations converged in different alarmist storylines on dangerous climate change: climate change as an external enemy/threat, climate conflict narratives and a tale of precaution. Let me shortly outline the core elements of these political storylines in more detail.

Storyline 1: Dangerous climate change as a threat for humanity

A first discursive storyline strongly featured by the alarmist camp revolves around the notion of climate change as an external threat to humanity. It serves as an *introduction* and sets out the very general plot of the alarmist version of climate change and security discourse. And in this plot climate change – as an anthropomorphized threat (see Chapter 4) – takes up the role of the *anti-hero*. Another actant role in the global climate security plot is *the victim*. Since climate change is a threat of planetary dimensions, this victim role is ascribed to humanity as a whole: “climate change is a security issue, but it is not a matter of narrow national security. It has a new dimension. It is about our collective security in a fragile and increasingly interdependent world” (Becket in UNSC 2007a: 19). Given the stalemate of international attempts to mitigate the climate risk, there is no actual *hero* in alarmist climate security discourse – there is only a potential one and this is the international community: “climate change is going to destroy our planet and our prosperity, if we don’t agree on a common approach” (*Süddeutsche Zeitung* Online 2010, author’s translation).

The plot of dangerous climate change (see e.g. Grove 2010: 539–543) is constructed upon several pairs of opposition that express particular normative judgments and define causal relations. A first of these binary pairs is the strict division between the natural and the social worlds. Be it climate change as a threat or enemy, be it climate change as a disease, or as an apocalypse – in all these cases climate change is represented as something that *comes upon us* and as something that has to be fought. The discursive effect is that the intrinsically social nature of 'the environment' becomes blurred (Swyngedouw 2010a: 222).

A second boundary is drawn between a 'normal' state of nature, i.e. the climate and related ecosystems in a stable equilibrium, and a 'dangerous' state of these systems (Boykoff *et al.* 2010: 58). This distinction has roots in the general climate change discourse, which constructed the two-degree Celsius (2°C) target as a threshold between a manageable and an uncontrollable version of climate change. The 2°C target is a highly arbitrary and hence intrinsically political demarcation (Lenton 2011: 451; Randalls 2010: 598–599). It has its roots in a mid-range emissions scenario of global warming in the second IPCC report in 1995, which was associated with an average global temperature increase of 2 degrees (Liverman 2009: 6). The 2°C benchmark gained political importance as the EU adopted it as a maximally tolerable amount of warming. The most prominent rationale for adopting this target is fear from large-scale discontinuities or irreversible extreme changes. These are, for example, a dieback of the Amazon rain forest, a change of Indian Monsoon and/or West-African Monsoon as well as the overturning of the North Atlantic Thermohaline Circulation (Lenton *et al.* 2008: 1788). Such irreversible major changes are discursively represented by the tipping point metaphor[4] – the notion that a hitherto gradual development, when overshooting a critical threshold, might become irreversible and uncontrollable. The tipping point concept is crucial for the climate change and security discourse in two ways. First, while in the IPCC Assessments +2°C is just a scenario which raises the probability of the occurrence of large scale discontinuities, throughout the climate security discourse +2°C becomes a tipping point in itself (between a gradual and an irreversible climate change). And in the climate security discourse this tipping point not only refers to natural changes but becomes linked with chaotic developments of social phenomena: "Much of the concern for the security implications of climate change relates to the possible consequences of large-scale and/or rapid disruptions to economies, societies and ecosystems" (UNGA 2009a: 15).

In sum, the plot of the dangerous climate change narrative resembles old and deeply ingrained fears of a nature gone wild: "No one can fight the wrath of nature" (Sorcar in UNSC 2007b: 10). Ever since the rise of Romanticism in the nineteenth century (succeeding a rather scientific treatment of nature in the European Enlightenment) and the first ecological catastrophes in the late nineteenth century, "nature would also be partly symbolized as dangerous and threatening, as fearful – yet man-made – in its urban manifestations" (Swyngedouw 2010b: 306). From the very beginning, such an ecological fear has been intermixed with anxieties from the poor, which, as we see below, also haunt the climate security discourse (Hulme 2008: 9–10).

Storyline 2: A tale of precaution

A second narrative that is shared by the alarmists in climate security discourse follows from the notion of nature out of control and the tipping point metaphor. It helps make sense of the temporal dimension of the climate threat and translates the assumptions on chaotic climate change into a political moral. In short, this storyline follows the *script of a tragedy*. There is a political subject – the *tragic hero* in this narrative – confronted with a problem or an enemy whose development or future actions are unknowable. The misjudgment of the political subject according to this narrative goes back to human nature (Taleb 2010: 8). As the subject deduces its predictions about future development from past experiences, she ends up underestimating the threat. In the end, the threat takes some unexpected turn and the *tragic hero* dies or at least suffers considerably (Methmann and Rothe 2012: 331–333). In climate change and security discourse this tragedy is often narrated through *analogies* to past civilizations or populations. Given their lack of experience with climatic changes, it is for example argued that the Mayan or the Egyptians were overwhelmed by a certain climatic shift (see e.g. Bojanowski 2011; Heinemann 2011). This ultimately led to their decline as civilizations (see e.g. Haroon in UNSC 2011b: 42).

In addition, articulations in climate security discourse often refer to strategic situations for the military:

> We never have 100 percent certainty [...] We never have it. If you wait until you have 100 percent certainty, something bad is going to happen on the battlefield. That's something we know. You have to act with incomplete information. You have to act based on the trend line. You have to act on your intuition sometimes.
>
> (UNGA 2009a: 10)

The human inability to imagine the unexpected has been made prominent through the *black swan* metaphor (Taleb 2010: xxi–xxii).[5] In security thinking, the black swan refers to extreme events with a low probability and a high magnitude. These are also labeled as *perfect storm* events, where a storm stands as a metaphor for all kinds of sudden shocks: "Climate change will raise the risk of the type of 'perfect storm' events seen in 2008 when food prices rocketed due to drought, high oil prices and trade restrictions" (Mabey 2011: 98). In addition, precautionary tales are organized around the motional collective symbols and metaphors of climate change as a race (Chapter 4). In their most extreme version, the precautionary storylines even foretell an apocalyptic catastrophe: "Climate change has become a major security issue that could lead to 'a world going up in flames,' the United Nations' top environment official has warned" (Bloomfield 2007). Other than normal dangers or catastrophes, climate change here represents the potential end point of a linear temporal development, as exemplified in the argument that "everyone's future is at stake now" (Aboud in UNSC 2007a: 35).

The moral of the precautionary narrative should have become clear: given the threat of climate change, the international community must act now – even in the absence of any concrete knowledge: "The key is to anticipate the need for action before tensions escalate to the point of no return" (Gorbachev and Severino 2007). At the same time one must always be prepared for the very worst – or the unexpected: "The mere possibility of such an unthinkable event is sufficient reason to seriously look for ways to prevent a future that none of us will be able to handle" (Sorcar in UNSC 2007a: 10). And political actors must not rely on past knowledge and experiences. A good example of this rationale is the analogy made by Al Gore in his Nobel Prize speech, comparing the current situation for the world community with 1939, when a majority of world leaders still would not see the danger posed by the German Nazi regime (Jordan 2007: 14).

Storyline 3: Climate conflict narrative

A final core storyline articulated by the alarmist faction is that of climate conflicts. In short the climate conflict narrative revolves around a neo-Malthusian script (Hartmann 2014: 770): "When crops fail, people may take up a gun to make a living" (Hsiang quoted in Carrington 2011: 23). The narrative is structured in *two episodes*. The first expresses humanitarian concerns for the victims of climate change, often drawing on a *human security* concept. At this point, the climate change and security discourse clearly overlaps with a broader discourse on climate change, responsibility and global justice (Phone interview with a New Delhi-based climate security researcher, December 9, 2011). The second episode, on the contrary, focuses on the secondary effects of climate change such as conflicts, migration or failing states as a source of insecurity.

These two episodes of the climate conflict narrative already indicate the actant structure of the plot. In the *first episode* there is no real anti-hero, but just a series of *subject-less risks*: the direct impacts of climate change such as sea level rise, droughts or natural disasters. The episode focuses on the figure of *the victim*. The victim is marked by its vulnerability against these direct impacts of climate change (Mason 2014: 808–809). The figure of the victim is furthermore marked by its *passivity* and *helplessness* (Bettini

2013). This is linguistically realized through the massive employment of passive constructions: "those displaced [...], those affected" (UNGA 2009a); "those most threatened" (Peesôa in UNGA 2009b: 20).

However, the second episode of the climate conflict narrative turns this actant structure upside down. Here the vulnerable and poor become themselves a source of danger (Oels 2013) – as climate-induced migrants, they become the *tragic heroes* or *anti-heroes* of the climate conflict narrative: "If there's no migration factor, if it's just local level conflicts between pastoralists and farmers nobody – quite frankly – really cares" (Phone interview with a UNEP employee, November 25, 2011). In contrast to the humanitarian tone of the first episode, the second part of the climate conflict narrative is basically constructed along the technical and physical collective symbols outlined in Chapter 4:

> "What we see in Darfur is an environmental change phenomenon unfolding that *puts pressure* on local communities," Mr Steiner said. "*Combine* that with potential *tensions* and you very quickly get a *potent mix* within which *increased pressure* can result in conflict." '
>
> (Steiner quoted in Bloomfield 2007, emphasis added)

Besides these technical metaphors, biology and health metaphors are particularly dominant in this episode, as exemplified in these quotes: "Scarcity breeds fear, which in turn fuels conflict" (Ogwu in UNSC 2011a: 10); "the spread of environmental migration" (Lambert in UNSC 2011b: 22). While the mechanistic and health metaphors seem to contradict the humanitarian tone in the first episode, they actually resonate well: in both framings, human agency and the subject disappears. The one produces a mechanistic picture of factors triggering other factors; the other constructs the vulnerable as passive victims.

Last, the underlying neo-Malthusian logic in such narratives means that climate conflicts are always located in the so-called developing world (Hartmann 2014: 774) – and particularly in Africa (Phone interview with a UNEP employee, November 25, 2011). As such, they can fuel anxieties from the dangerous poor that have a longer tradition in occidental discourses.[6] Alarmist climate security discourse resembles old colonial stereotypes of 'Third World' peasants degrading their environment by over-farming and overgrazing due to population pressures. Betsy Hartmann has called this a *degradation narrative*:

> The ensuing soil depletion and desertification then lead them to migrate elsewhere as "environmental refugees," either to other ecologically vulnerable rural areas where the vicious cycle is once again set in motion or to cities where they strain scarce resources and become a primary source of political instability.
>
> (Hartmann 2013: 92)

In colonial contexts such narratives of protecting local populations were also used to legitimize political interventions by imperial powers. The moral of the narrative in climate security discourse is quite similar: supporting the victims of climate change is a both a moral obligation and a national interest for the industrialized countries.

Climate security rationalists and climate security skeptics

In addition to the alarmists, there are the rather cautious voices that can be summarized as a faction of 'rationalists.' These articulate a rather scientific framing of climate change and security linkages and in general stress the *multi-causality* and *complexity* of climate–security linkages. This applies to most of the academic debate on climate change and security and also to research oriented think tanks. My interviews have shown that many bureaucrats dealing with the issue are also somewhat cautious when it comes to climate change and security linkages: "I don't think it's so simplistic" (Phone interview with a member of DG Humanitarian Aid, September 27, 2011). Such actors have not produced a genuine set of storylines but rather relate to the alarmist storylines just outlined in a cautious and critical fashion. For example, climate conflict storylines are not generally rejected but added with a cautious note to account for the multi-causalities and complexity of any conflict (see Ide and Scheffran 2014: 2–3). Last, there are those voices that stand in opposition to the alarmist storylines – with the faction of *climate security skeptics*.[7] These have attacked climate security storylines outlined as a Western discourse blurring the industrialized countries' responsibility for climate change and shifting the focus to the global South (Boas 2014: 152–153). As an alternative to the security storylines, emerging powers have repeatedly articulated climate change as a development issue. In addition to representatives of bigger emerging and developing countries, some academics joined the camp of climate security skeptics by warning against the negative implications of a securitization of climate change (Hartmann 2010). Table 5.2 summarizes the different factions that together made up the discourse coalition on climate security in the peak phase of the discourse in the late 2000s.

By way of conclusion, one can say that, on the one hand, the main explanation for the success and dominance of the climate security discourse is the massive expansion of the discourse coalition after 2007, with the inclusion of authoritative voices (climate celebrities) as well as critical actors. At the same time, the coalition is marked by *inner contradictions* and competing framings and storylines. The next section outlines the discursive strategies that have enabled both the growth of the CSDC and the pacification of the discursive field through the construction of consensus.

Table 5.2 The transnational CSDC in the late 2000s

Factions	*Subject positions*	*Storylines*
Alarmists	Security think tanks Climate celebrities Forerunners Victims NGOs	Dangerous climate change as a threat for humanity Tales of precaution: a looming apocalypse Climate conflict narrative
Rationalists	Academics Bureaucrats Science oriented think tanks	Redefinition of security Complexity and multi-causality of climate change security implications
Climate security skeptics	Emerging powers Academics	Climate change is a development and not a security issue Negative impacts of securitization

How climate security became a floating signifier

A common orientation towards climate change and security linkages is what constituted the CSDC. Hence, the primary discursive strategy that was a necessary condition for the CSDC was the emancipation of climate change from a broader discourse on environmental security in the first years of the twenty-first century. From a discourse theoretical perspective, what happened here was the *inversion* of the representational relations between climate change, the environment and security. Climate change was isolated from a chain of equivalence, which subsumed it under global environmental problems alongside other concerns such as deforestation or air pollution. Climate change became an *emblematic issue* (Hajer 1995: 20) of environmental security, or nodal point that structured its own discursive field – climate change and security discourse. A growing number of other environmental problems were now seen as either symptoms or consequences of a warming climate: "It would appear that all existing and potential threats from environmental damage come together in the overarching threat posed by climate change" (Wenaweser in UNSC 2007a: 28–29).[8] This inversion of the representational structure was crucial as it constituted the basis of the further discursive strategies that followed.

With the inversion process described above, the *climate security signifier* was born as a novel *nodal point* in international climate change as well as in security discourses. In the following years, then, and especially after the discourse gained momentum in 2007, the meaning of this signifier was fiercely contested. This contestation was played out on the basis of different storylines. First, there were *storylines* of the alarmist faction that articulated climate change in a *military security* context – mainly from a perspective of Western industrialized countries. Climate change was brought into an equivalential relation with other globalized security risks such as "nuclear proliferation or terrorism" (MacFarquhar 2011). Such articulations present climate change *itself* as a threat to national or international security (Phone interview with a US climate policy expert, September 16, 2011). Such an argument has mainly been brought forward by representatives of industrialized countries (forerunners) as well as climate celebrities. A *second articulation* from a narrow understanding of security constructs climate change as a sort of *super-determinant* of global security: "climate change is going to be the security issue of the 21st century" (Keelty quoted in Walker and Faulkner 2007); "Global warming's impacts on natural resources and climate systems may create the fiercest battle our world has ever seen" (Knickerbocker 2007). These articulations isolate climate change from an equivalential chain with other security threats by stressing its exceptionality and presenting it as a hypervariable: "The collapse and chaos associated with extreme climate change futures would destabilize virtually every aspect of modern life" (Campbell *et al.* 2007: 9). One can find such a framing in all articulations from the camp of the alarmists.

A *third articulation* of climate security was mainly made by the subject position of the victims. In particular, SIDS have referred to the climate security signifier in terms of *national survival* (Interview with a German governmental advisor, Berlin, September 14, 2012), as expressed in this example: "For small island states, the security implications of climate change have been, first and foremost, a threat to our very existence as sovereign nations" (Vunibobo in UNGA 2009b: 15). Such articulations construct a chain of equivalence between climate change security, sea level rise, cultural and collective identity, and statelessness: "Some of our Pacific countries [...] will cease to be states" (Moses in UNGA 2009b: 2). While the three framings present different referent objects, their common denominator is the representation of climate security in terms of *survival, exceptionalism and emergency.*

In opposition to these three articulations are framings of the climate change threat in terms of *human security*, strongly promoted by, for example, NGOs and the actors that see themselves as victims of climate change: "Climate Change has helped us all understand that security and prosperity cannot be nation-centric and must be at the level of the individual and hence we must now revisit security in terms of human security" (Joshi in *Hindustan Times* 2009). Climate change became related to development and humanitarian problems such as "food insecurity, increased poverty and reverse development" (Salami in UNSC 2011a: 12). The security signifier was taken out of its conventionalized military context:

> Security policies exclusively based on national sovereignty appear less and less appropriate in this context. To resolve the climate/energy-security dilemma means that we need to rethink thoroughly the scope of our policies using a broader concept of security."
>
> (Verbeke in UNSC 2007a: 5–6)

While the storylines presented so far fight over *the right* representation of climate change and security, there are some articulations that oppose a climate security framing in general. In the previous section, I called the respective subject positions *climate security skeptics*. Basically, such articulations construct an antagonism between a security framing on the one hand and a proper climate change discourse on the other, thereby externalizing climate security to an illegitimate discursive outside:

> There is a clear link between climate change and development, established by sound scientific research. There is also a clear link between development and security. There is no direct linkage, however, between climate change and security.
>
> (Peesôa in UNGA 2009b: 20)

Climate change is not articulated alongside demands for security, but rather, in a discursive context structured by the prominent signifier of *sustainable development* including demands such as economic growth, welfare, international equality, etc. What we can observe here is a discursive struggle around the proper framing of climate change. The result of this struggle in terms of discourse theory is that climate security became a *floating signifier* (see Chapter 3). The climate security signifier became characterized by its *over-determination*, marked by a surplus of meaning, which ultimately could not be fixed. As long as there was this discursive struggle, however, no hegemonic CSDC could develop. What was needed was a fixation of meaning through the introduction of different *nodal points*, which could pacify the battlefield of the climate security discourse.

How the climate security discourse became stabilized

Nodal points in the climate change and security discourse refer to *important signifiers and collective symbols* that allow the competing versions of climate security to be articulated as a coherent whole. This production of consensus went so far that even climate security skeptics could warm to such consensual narratives. The nodal points hence partially stabilized the discursive field so that a hegemonic representation of climate security could develop.

The first of these nodal points is the concept of climate change as a *threat multiplier* that turned up in the influential report by the CNA on the national security implications (Center for Naval Analysis 2007: 6) and was brought into the UNSC debate through the British report written in preparation for the meeting (see UK Mission 2007). It is commonly cited in every report on the issue and often presented as a *consensual concept*: "Today, we generally regard climate change as a threat multiplier" (Štiglic in UNSC 2011b: 2). Crucial to the threat multiplier metaphor is its ambiguity and vagueness: to act as a multiplier could mean either to amplify single factors or to increase the total number of threats. And the metaphor leaves entirely open how this multiplication of threats actually happens. The catachresis[9] of a threat multiplier is important in climate security discourse, as it resolves contradictory takes on climate change conflicts by stating that climate change does *not pose* a security threat *itself*. But climate change can retain its central position in the security storyline, as it worsens a whole series of factors that are commonly associated with conflicts (Phone interview with a member of the UK FCO, December 15, 2011). This makes the entire variety of different security concepts that can be found in the discourse compatible: the threat multiplier concept repeats the assumptions that climate change threatens the vulnerable and multiplies the risks that derive from their inability to cope with these threats (Phone interview with a UNEP employee, November 25, 2011). Finally, the "future-conditional tense" (Baldwin 2014: 213) of the threat-multiplier metaphor makes climate security arguments resistant against critique based on past empirical evidence. Even if empirical case studies did not find robust correlations between climate change and conflict in the past "the future is yet another country – an as of yet unoccupied political space" (Hartmann 2014: 772).

A second nodal point in climate security discourse is the concept of *climate refugees* and synonymic signifiers such as climate-induced migration (Phone interview with a Geneva-based climate security researcher, September 13, 2011). The figure of the climate migrant as a nodal point is important in several ways. First of all, CIM is a central and major issue across the whole discourse. Second, it is a crosscutting issue, as it concerns very different policy fields such as security, environment, development policy as well as international law – it hence bridges different discourses. Third, it is a consensus-producing issue, as it resonates with both narrow, military security thinking *and* humanitarian concerns. This also means that the concern regarding refugees opened the conceptual door for further actors to enter the discourse coalition. Let me give a few examples: in 2009 a study by a humanitarian NGO and the UN University warned that climate change-related grievances and problems in the middle of the twenty-first century could produce waves of migration on an unprecedented scale (Spiegel Online 2009). Such dramatic warnings were substantiated by estimations of the number of climate refugees reaching 200 million by 2050 (Spiegel Online 2009). Even if those warnings followed a rather humanitarian discourse, they cohered with narrow security rationales of European, Australian and US politicians: "The UN predicts that there will be millions of 'environmental' migrants by 2020 [...] Europe must expect substantially increased migratory pressure" (Solana 2008: 4).

A third means to produce consensus about climate change and security linkages, which hitherto has not been discussed in the literature, refers to what I would call *the force of example* through *emblematic cases*. To substantiate or authorize their articulations, speakers often draw on especially colorful or striking cases, which are emblematic symbols or prototypes of broader developments.[10] A first of such synecdoche is *the fate of small island states* existentially threatened by climate change (Interview with a German governmental advisor, Berlin September 14, 2011). SIDS are the paradigmatic prototype of the victim figure in many climate security storylines. They are taken as a symbol that stands in lieu of the fate of the whole planet: "We will all suffer sooner or later. No country is immune to the impacts of climate change, and the smallest of the SIDS will be the first" (Moses in UNGA 2009b: 2).

Another prominent example in this respect is the Darfur conflict (Hartmann 2014: 770). Darfur is taken as a symbol of the climate change and conflict storyline (Phone interview with an Oslo-based climate security researcher, September 29, 2011). The *pars pro toto* logic behind the storyline becomes clear in the following quote: " 'The world was already experiencing its first war partly caused by climate change [...]. The situation that emerged in Darfur will emerge in other parts of the world.' He [Achim Steiner] warned of a 'world going up in flames' " (Steiner quoted in Bloomfield 2007). Referring to such a powerful example as Darfur adds evidence and authority to climate conflict storylines and is for example quoted in the 2014 IPCC report on *Impacts, Adaptation, and Vulnerability* (IPCC 2014: 773). A third prominent example is the Arab Spring, which has been partially ascribed to climate impacts in North Africa and the Middle East and their negative impacts on food security in the region (Johnstone and Mazo 2011). A final prominent example is the reference to the Islamic State in Syria and Iraq: "Climate change and water shortages may have triggered the drought that caused farmers to relocate to Syrian cities and triggered situations where youth were more susceptible to joining extremist groups" (King quoted in Davenport 2014). By using the symbolic power of major international security crises, such articulations are able to link climate change and security even in the *absence of clear empirical evidence* for such linkages.

Finally, nodal points in climate security discourse are not reduced to linguistic symbols. As argued in Chapter 3 material things can also fulfill this function by becoming *boundary objects* that bridge different discourses. A good example for such a boundary object in climate change and security discourse is the security *fence* at the border between *India and Bangladesh* (see Figure 5.1).

To understand why this can function as a representation of climate security, note that, since the first concerns over environmental migration in the 1990s, Bangladesh has been represented as potential hotspot (e.g. Bloomfield 2007). A majority of the 150 million inhabitants of Bangladesh is living in the low-lying coastlands or on the banks of its 800 rivers: "Nowhere is the potential threat from climate change more worrisome than in Bangladesh" (Nayak 2011). And given the fragility of the country, it is feared, Bangladesh will not be able to cope with these streams internally, which means that they would spill into neighboring India (Chelleney 2008).

Figure 5.1 Security fence at the border between India and Bangladesh (reprinted with kind permission of Prashant Panjiar, copyright 2015).

The fence that would protect the Indian border from Bangladeshi immigrants has become a part of the climate refugee storyline:

> India has already started building a wall to stop Bangladeshis coming across. The predicted half-a-meter sea level rise means 34 million people not being able to stay where they are now. Where will they go? They will break through the boundaries.
>
> (Steiner quoted in Bloomfield 2007)

And in a fictional scenario of the situation in 2020 the German Advisory Council on Global Change (WBGU) anticipates that: "Many of these migrants cross the border illegally, often with the aid of people smugglers who know how to get through the border fence erected by India. Many people die in the attempt" (WBGU 2009: 123). Paradoxically, Indian representatives do not see any link between the security wall and the threat of climate change: "the primary message was that migration from Bangladesh towards India is not related to environmental and climate change factors" (Boas 2014: 154). And still the photograph of the fence circulates through Western media and think tank reports as a symbol of climate security and a warning of "what will happen to us" if we do not stop climate change. The fence becomes charged with a *performative* power, which is irreducible to the intentions and interpretations of its constructors.

A last powerful example is the case of *conflict over resources in the Arctic* that have become available due to the melting Arctic ice. Different articulations have predicted a future struggle or race for the natural resources of a melting Arctic between the surrounding countries Norway, Russia, Denmark, Canada and the US (*The Japan Times* 2009). The Arctic is hence articulated as an example of the storyline that conflict can be fuelled not only by natural resource scarcity but also the availability of key (fossil) resources. Again there is an *emblematic symbol*, which represents this storyline. And this a *Russian flag* placed into the Arctic seabed by a mini-submarine in 2007.

This placement was meant to express a territorial claim for the natural resources that are assumed to be covered by the Arctic ice (Ryan 2011). The act was live-streamed on the internet and was thus visible to the whole world. It is important to note that this has been a solely *performative* action without having any legal implications whatsoever.[11] But the message behind this act was clearly recognized by competitors, which were startled by the Russian move.[12] Contrary to other storylines in climate security discourse which do not present a clear-cut enemy figure, in storylines on conflict over the Arctic there is a *process of Othering* that constructs Russia as a political threat: "Russians are already becoming more assertive, bordering on aggressive" (Rodgers 2010). The Russian flag has become a symbol of the looming resource conflict and of Russia's political responsibility for having started this struggle: "Perhaps it would have been more appropriate for the Russians to have waved a green flag, because the act symbolically started the race to claim the oil beneath Arctic ice" (Sappenfield 2007).

How the signifier of climate security became emptied

The pacification of the discursive field of climate security did not take place by way of a clear-cut victory of one discursive project over all the others. What we notice, on the contrary, was how the initial alarmist faction in the CSDC revolving around a rather narrow security concept subsequently integrated contradicting and oppositional articulations. For example, it integrated the concept of human security and reinterpreted it as a source of potential conflicts, thereby relating it to international security. The climate security signifier in this sense acts like an enormous conceptual *sponge*, absorbing more and more concepts and issues (Phone interview with a UK climate security researcher, November 2, 2011). The discursive strategies at play in this process and the nodal points they produced are summarized in Table 5.3.

These strategies were able to partly harmonize the competing storylines on climate change and security. For example, even climate security skeptics such as India or Russia in the latest UNSC debate in 2011 agreed that climate change affects the security of SIDS and developing countries and

Table 5.3 Discursive strategies in the hegemonization of climate security

Discursive strategies	*Explanation*	*Effect*
Carbonification	Other environmental problems were reframed as climate change impacts	Constitution of climate change security as a genuine discourse object
Threat multiplier concept	Harmonization of competing articulations of climate security	Pacification of the discursive struggle
Climate refugees	Discourse on climate-induced migration as a major issue across the whole sample	Introduced novel subject positions into the CSDC (e.g. humanitarian NGOs) Brought different voices together as a cross-cutting issue
The force of example	Paradigmatic cases such as the SIDS, Darfur, the Arctic, or Bangladesh are taken as emblematic symbols	More and more storylines of climate change and security = the climate security signifier becomes emptied

that it might act as a threat multiplier in the future. In its 2014 assessment report, the IPCC working group on *Impacts, Adaptation, and Vulnerability* discussed the issues of climate-induced migration, climate change as a cause of conflict, and the geopolitical implications of climate change for the first time (IPCC 2014: 766–776). This is particularly remarkable because in all its previous reports the IPCC has been reluctant to touch these issues due to scientific uncertainties and their controversial political nature. The IPCC 2014 report is hence a perfect example of a growing hegemonic consent on the security implications of climate change.

As a reaction to contestation, climate security was continually rearticulated and thereby related to ever more phenomena through the practices and storylines outlined in this chapter (Phone interview with an Oslo-based climate security researcher, September 29, 2011). Climate security refers to conflicts such as Darfur, it is visible in a struggle for the Arctic, it refers to the well-being of people in a place like Mali, becomes linked to energy insecurity, is related to the spread of terrorism, etc. One could probably continue this list over the next two pages. This resulted, on the one hand, in a very high degree of *discourse structuration*: at the end of the 2000s, one could not reasonably deny anymore that climate change, in some way or another, could have security implications. Climate security as a signifier represented a growing number of completely different referent objects – see the following example: "climate change threatens markets, economies and development gains. It can deplete food and water supplies, provoke conflict and migration, destabilize fragile societies and even topple governments" (UNGA 2009a: 6).

Climate security represents a paradigmatic example of how a hegemonic discourse is established through the creation of antagonism and an empty signifier (see Figure 5.2). In hegemonic climate security discourse, the external other is represented by *dangerous climate change.* Dangerous climate change represents a whole range of different threats and dangers. *Climate security* as the negation of this threat at the same time represents an ever-growing number of political demands related to these challenges and problems. As an empty signifier it temporarily fixes meaning in climate security discourse. At the same time, it becomes conceptually conflated, or *over-determined* in the terminology of discourse theory. This means it becomes ever more vague – or emptied (Interview with a climate security researcher, Berlin September 14, 2011).

Political implications in global climate governance

There are two particular arguments for why the Copenhagen School's perspective misses the full implications of a securitization process in the case of climate change. First, according to the Copenhagen School there is a strict separation between the securitization process (the speech acts) and its political effects (the exceptional measures that are legitimized). This means that all the political practices that happen in the process of a discursive struggle, such as practices of knowledge-production on climate conflict linkages, cannot be seen. Second, and related to this, the Copenhagen School conceptualizes securitization as a transfer from one discourse or policy field – security – to another (the securitized issue area). This makes it blind to all the mutual interferences between discursive fields – concepts and practices traveling from one field into the other and

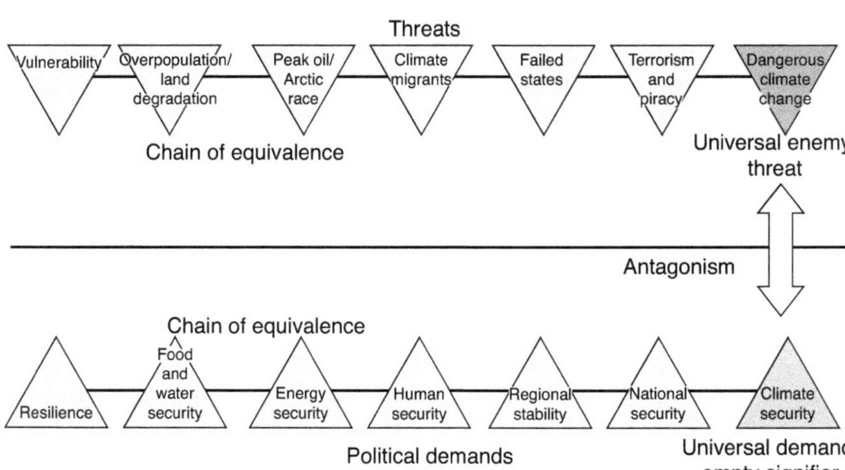

Figure 5.2 Climate security as an empty signifier (source: author).

back, assemblages that are established across the different regimes, or the evolution of new forms of government transcending the borders between different policy fields. In the remainder of the chapter, I show that the climate change and security discourse does not constitute a governance object on its own, but up until now has mainly affected *international climate governance*. First, it fostered a shift towards *resilience* as a new paradigm in global climate governance. Second, following from that, it has helped stabilize a *technocratic, market-based* and *post-political approach* to climate governance.

From mitigating climate change to promoting resilience

The securitization of climate change can be understood as "recombinatorial processes through which techniques and technologies [in global climate governance, author's note] are reworked and redeployed" (Collier 2009: 91). A security framing of climate change did not succeed or replace an economically or ecologically oriented problematization of climate change. Security has not become *the* hegemonic representation of climate change (Phone interview with a US climate policy expert, September 16, 2011). But it has been articulated and has had an influence in nearly all the fields and niches of the governmental landscape of international climate governance. Yet these political implications only become visible if we acknowledge that implications are already to be found in the discursive struggle itself. By taking up such a perspective, one can spot many different places at which climate security links up with existing practices in the climate governance problem space. The heterogeneity of these sites is owed to the contingency of the climate security discourse and the related vagueness of the climate security signifier.

On the one hand, one can clearly say that the climate security storylines presented in this chapter resonate with a global management approach to climate governance. Storylines that paint climate change as an enemy of humanity, as well as related security concepts such as collective security, *reinforce a global framing* of climate change. On the other hand, a security take on climate change stresses the importance of assessing, monitoring and predicting possible climate risks. It hence supports the extension of a global system of carbon *monitoring and control*: a comprehensive global carbon surveillance system – based on the UNFCCC's measuring and reporting mechanisms, climate models and satellite based surveillance (Bulkeley *et al.* 2011: 20–21). As Simon Dalby notes, a key actor in this global monitoring of the atmosphere and related earth systems has been the military – due to the concentration of remote sensing capabilities in the hands of militaries, attempts to weaponize regional weather modification or the debate on a Nuclear Winter (Dalby 2013: 40–42). It is in this need for *big climate data* that a geopolitical security take on climate change converges with neoliberal, market-driven forms of governance that dominate contemporary climate governance:

> The planet is now not only a matter for measurement as an object to be known through the practices of remote sensing, but much more obviously now a matter of calculation and computation wherein ecological phenomena are divided up, surveyed, demarcated and marketed, bought and sold in the complicated computations of carbon markets.
> (Dalby 2013: 44)

The primary approach to secure climate risks in the international climate regime draws on market-based emission offsetting schemes including the Clean Development Mechanism (CDM) or the REDD+ mechanism (Reducing Emissions from Avoided Deforestation and Land Degradation). Pointing to the intimate relationship between market-based mitigation measures and climate security has two implications. On the one hand, it shows that security policies do not necessarily break with existing rules – such as the dominance of neoliberal policies – which is other than suggested by the Copenhagen School. In fact, "tools to pursue security often take market-based forms" (Dalby 2015: 8). On the other hand, the convergence of security and finance points towards the often-violent nature of market mechanisms such as the CDM. This violence includes the forceful nature of carbon offsetting, reforestation or biofuel production in Africa projects and Latin America. Such projects often include practices of land grabbing and require military violence for their protection (Dunlap and Fairhead 2014: 941–942).

Moreover, the convergence of climate change and security thinking also implies a shift in the methods of anticipating dangerous climate futures. The basis of a risk management approach to climate change was a range of different technologies to *calculate* an insecure future (Anderson 2010: 784), such as stochastic modeling or probability assessment. Climate security discourse problematizes such a classical risk management approach, as the latter systematically underrated the threat of low-probability-high-impact risks and non-linear developments (Mabey 2008: 4). Due to the perceived complexity and incalculability of climate security risks (Grove 2010: 545), a securitization of climate change promotes a more imaginative or creative approach to anticipate future climate risks (de Goede and Randalls 2009). Such creative practices of creating *future-conditional knowledge* include the development of scenarios: "Scenarios are meant to help organizations handle uncertainty of the future, as a way of preparing for what would otherwise be unexpected crises" (Maas and Tänzler 2009: 16). Other practices of performing future climate threats are role plays or preparedness exercises in which participants are embodying an 'as-if' future (Anderson 2010: 787).

Finally, a securitization of climate change supports a resilience and adaptation approach to the problem of climate change (Phone interview with a member of DG Humanitarian Aid, September 27, 2011). A resilience approach focusing on adaptation to climate change impacts in vulnerable regions seeks to secure the failures of a market-based mitigation of climate risks (Corry 2014; Methmann and Oels 2015). Emanating from ecosystem research, resilience refers to the capability of a system to quickly recover after sudden shocks (like major disasters) without necessarily moving back to its previous state (Walker and Cooper 2011: 146). The ability to recover requires being flexible and adaptive (Reid 2012: 71). In Western security discourse, resilience constitutes the foundation of a post-interventionist paradigm: "in a post-interventionist/resilience paradigm, the emphasis is on prevention rather than intervention, empowerment rather than protection, and work upon the vulnerable rather than upon victims" (Chandler 2012: 216). It is precisely this rationale of *empowering the vulnerable* to increase their resilience that informs the contemporary climate security discourse and its take on adaptation: "Adaptation requires empowering people, building their resilience" (UNGA 2009a: 6); "hope lies in enabling the vulnerable to build resilience in order to avoid humanitarian catastrophes" (Pessôa in UNGA 2009b: 21).

According to David Chandler (2014: 55), while resilience shares with neoliberalism its distrust in central government and planning, it was actually popularized as a solution to the intrinsic shortcomings of neoliberalism. The problem was that while neoliberalism drew on a radical critique of government intervention, informed by the notion of the radical complexity of systems such as national economies, it could not translate this radical critique into a program of governance. The reason was that for markets to be created a considerable amount of state intervention and social engineering was necessary (Schmidt 2014: 4). The creation of markets for carbon offsets is a perfect case in point as these required massive amounts of state intervention: from the scientific efforts to calculate and measure carbon emissions through the institutional setting for CO_2-certificate trading, to the use of military force to secure CDM projects in the global South. For neoliberalism, such interventions pose serious governance obstacles, given the assumption that knowledge for such interventions is necessarily incomplete. In global climate governance, these problems become apparent in the failure of offsetting schemes to reduce any kind of future climate risks. As interventionist policies, carbon markets and similar mitigation measures undermine their own effectiveness.

Resilience thinking, on the contrary, embraces complexity as a governmental paradigm. It promotes local ownership, self-reflexivity, and stakeholder-participation to overcome both market – as well as governance – failures (Chandler 2014: 57). Vulnerable communities are conceptualized in terms of resilient systems that have developed particular kinds of tacit knowledge and adaptation skills, which are passed on from generation to generation. Governmental technologies of following an ethos of resilience seek to mobilize exactly these localized forms of tacit knowledge for their own governmental agendas. While it has been feared that a securitization of climate change could involve an increase in humanitarian interventions in potential climate hotspots (Hartmann 2010: 241) at the moment an opposite trend is observable. In line with a *post-interventionist approach*, adaptation policies try to increase the preparedness of the vulnerable from a distance so that these do not become victims of dangerous climate change and subsequent interventions become obsolete (Corry 2014).

At COP 16 in Cancun 2010, the UNFCCC members decided to set up an *adaptation framework* to provide "new and additional, predictable and adequate funding [...] taking into account the urgent and immediate needs of developing countries that are particularly vulnerable to the adverse effects of climate change" (UNFCCC 2010a). They furthermore agreed on a Green Climate Fund as a major international funding instrument to support developing countries, launched one year later at COP 17 in Durban (decision 3/CP.17). To oversee the various adaptation activities under the UNFCCC umbrella, an Adaptation Committee was established in 2012 (UNFCCC 2012). The UN Climate Summit in September 2014 in New York also put a special focus on resilience, since "Strengthening resilience to the changing climate is essential" (Ban Ki-Moon quoted in United Nations 2014: 1). As a result a group of developed countries committed themselves to channel US $18 billion to the developing countries during 2014–2020 in support of adaptation to climate change impacts. It is certainly not possible to discuss the whole UNFCCC adaptation framework in detail here. What is important to note, however, is that this framework is absolutely in line with a post-interventionist paradigm of resilience, which shifts responsibility to the vulnerable themselves. This clearly mirrors forms of government at a distance, which have been associated with advanced liberal government (Joseph 2013). Least developed and vulnerable developing countries are required to identify their own vulnerabilities by preparing so called National Adaptation Programs (UNFCCC 2010b: 5–6). The responsibility for adaptation and resilience projects is shifted to national and local governments, while the UNFCCC restricts itself to providing financial and technical assistance as well as knowledge resources including, for example, methods of risk and vulnerability assessment.

While it is wrong to ascribe this shifting focus from mitigation to adaptation directly to a security framing of climate change, it certainly promoted a more prominent role for adaptation as a part of climate resilience (Chandler 2014; Corry 2014; Methmann and Oels 2015). The reason is that at the heart of resilience thinking is the assumption that a certain degree of risk and danger is inherent in any social or natural system and hence inevitable. In a world of omnipresent danger there is a need for continuous adaptation as a way to learn from past events and to prepare for uncertain future ones (Evans and Reid 2013: 91). Articulations in the discourse on climate change and security ultimately support exactly such a picture of omnipresent danger by presenting some degree of dangerous climate change as inevitable. At the same time, resilience revolves around the notion of plexity and non-linearity. This perfectly resonates with the discourse on non-linear climate change and the limits to knowledge it poses. Stressing complexity and context-dependency resilience thinking not only resonates with the voices of alarmists but also with the more cautious voices outlined above. In other words, even if the security implications of climate change are unknowable, it still makes sense to invest in the resilience of the vulnerable as a means to be prepared for all different sorts of danger.

Securitization as post-political populism

What is notable about the politics of neoliberal security in mitigation as well as the paradigm of resilience in adaptation governance is that they would hardly qualify as exceptional measures in the sense of the Copenhagen School. On the contrary, the instruments on adaptation and resilience under the UNFCCC umbrella could be qualified as rather piecemeal, technocratic, and often messy forms of risk management.[13]

Recall what I said about the intimate relationship between security and risk in the theory chapter of this book (Chapter 3). The construction of an antagonism is a necessary condition for any risk management practice: it constitutes the social space which then can be governed according to a logic of risk. A (macro-)securitization of climate change fulfills exactly this function: it articulates global warming as an external threat that coincides with the limits of the planet and thereby constitutes the latter as a homogeneous social space (Methmann and Rothe 2012: 329). What became apparent in this chapter was that, despite all their differences, the manifold articulations of climate security all converged in a particular image of the climate crisis as a planetary emergency and of climate change as *a threat to humanity*. All the different threat constructions – the direct impacts felt by the vulnerable, as well as second order threats for developed countries – are subject to this meta-framing. In line with the Copenhagen School, one could argue that climate change represents a macro-securitization at the global level (Buzan and Wæver 2009: 258).

The particular macro-securitization that permeates the whole discourse of climate change and security is the main reason why a securitization of climate change does not come along with the adoption of any exceptional measures in the sense of the Copenhagen School. Rather, the macro-securitization strengthens existing technocratic practices of global carbon control and monitoring as well as the politics of resilience that both characterize international climate governance. First, no country can deal with this overarching threat alone – hence there is no alternative to a global climate deal under the umbrella of the UNFCCC. Second, humanity's war against dangerous climate change is a *war of all against nothing* as "we are all in the same boat" (Pachauri quoted in Spranz 2007, author's translation). This particular invocation and similar ones of the collective subjectivity of humanity ('we') has a clear depoliticizing effect. All antagonisms, which characterize the international order in general, and global climate politics in particular, become blurred against the background of the overarching threat of climate change. And third, the complexity and overwhelming nature of the climate change threat simply makes extreme measures (particularly in adaptation) inappropriate. As suggested by the radical critique of abstract, centralized knowledge of resilience thinking – any military apparatus would be overwhelmed by managing the social implications of climate change, and hence the only way to deal with such second order security threats from climate change is to increase the resilience of the communities at risk (Methmann and Rothe 2013).

The climate security discourse represents a paradigmatic example of post-political populism, which seeks to secure the currently hegemonic international order (Swyngedouw 2010a). This logic of depoliticization is outlined in Figure 5.3.

The only appropriate weapons in this fight, according to the discourse, are hence the technocratic UNFCCC mechanisms to manage mitigation and adaptation risks: "mitigation and adaptation are two sides of the same risk management coin" (Webster in UNSC 2011b: 11). Alternatives to such

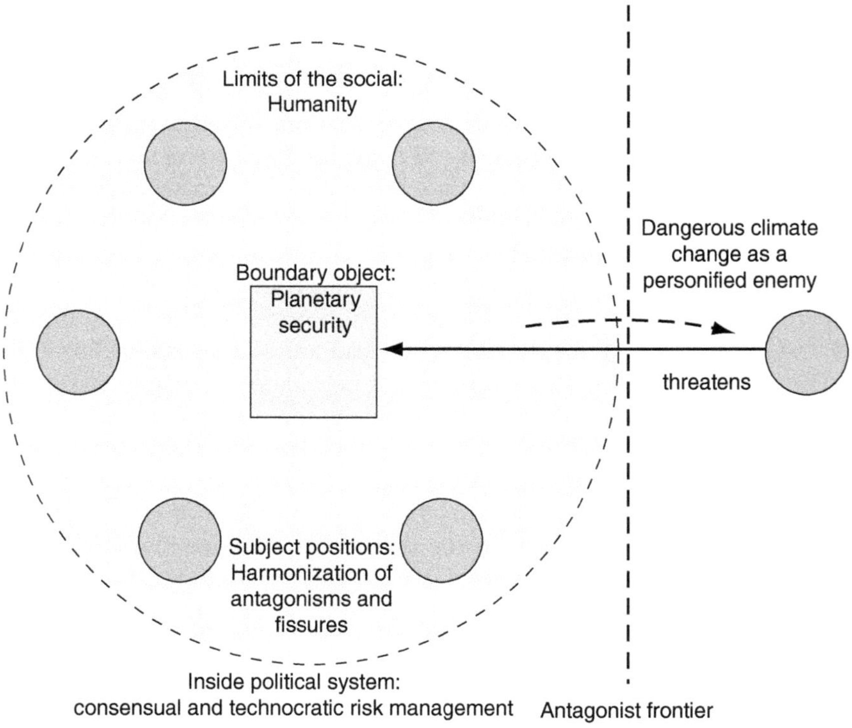

Figure 5.3 Climate resilience as post-political populism (source: author).

a technocratic risk management are *eradicated from the problem-space* of international climate governance. By outlining a catastrophic counter-factual of a global disorder in the future, climate security discourse furthermore romanticizes the current international order, which serves as a positive referent point. A populist logic is also apparent in the imagination of "the people know best" (Swyngedouw 2010a: 223), which in the climate security domain is represented by the focus on communities' resilience, stakeholder participation and tacit knowledge. Lastly, post-political populism draws on the vagueness of empty signifiers that do not refer to clearly identifiable governmental actions but are rather abstract and blurry concepts – as are resilience (Brassett and Vaughan-Williams 2015: 38–39), dangerous CO_2 (Swyngedouw 2010a) and climate stabilization (Boykoff *et al.* 2010). In sum, post-political populism defends the status quo by taking political questions out of the terrain of political contestation and subjecting them to consensual forms of political management.

Notes

1 An article by Barnett, initially written as a working paper in 2001, on "Security and climate change" (Barnett 2003) was one of the first studies that systematically and explicitly focused on the issue – further studies followed (e.g. Ashton and Burke 2004; Lacy 2005; Stripple 2005).

2 The perceived neutrality is what characterizes these political subjects, as they are either members of major international institutions or do not hold political office at all. They are further associated with a high moral authority and have a particular influence due to their institutionalized position. Due to these discursive attributions, climate celebrities represent particularly strong discourse multipliers.

3 However, there are also exceptions (e.g. Burke *et al.* 2009; Fetzek and Mazo 2014; Reuveny 2007).

4 The tipping point metaphor stems originally from epidemic research. In the early 1990s this metaphor spilled over to the discourse on social control and in particular on gang violence as part of a 'public health approach' to crime (see Gladwell 1996). The concept was then introduced into the emerging climate impact literature, referring to natural climate changes (see Lenton *et al.* 2008).

5 The black swan is an old metaphor to denote the improbable: based on the fact that all historical records of swans reported them as white, the existence of a black swan appeared as impossible.

6 Lohmann shows through a genealogy of Malthusian thinking that the latter has been one of the major threat narratives in Western discourses since early industrialization. It was fuelled by existential fears of capitalists and middle classes about the dangerous poor who by far outnumbered them (Lohmann 2005: 81).

7 One could also argue that these oppositional voices are not themselves a part of the climate security discourse coalition. However, I would argue that as long as they are oriented towards climate security storylines, even if they criticize them, they clearly represent a part of this coalition.

8 Ayşem Mert has called this discursive strategy a *carbonification* of environmental discourses, meaning that problems like deforestation or land degradation and air pollution were increasingly rearticulated in relation to climate change (Mert 2013).

9 A catachresis is a figurative term for which there is no literal expression. Hence, the catachresis constitutes a previously non-existent object by naming it (Howarth and Griggs 2008: 32–34).

10 The force of example, therefore, comes close to Hajer's notion of emblematic issues (Hajer 1995: 20) and to the rhetorical trope of synecdoche (a part represents the whole). Both refer to discursive strategies in which a single case is taken as representative for a broader phenomenon to which it is related. Both follow a logic of substitution: case X represents the broader phenomenon Y (Nonhoff 2007: 168).

11 Claims for the Arctic can be made in line with the United Nations Convention on the Law of the Sea (UNCLOS) on the ground of scientific evidence for the size of the continental shelf under the Arctic sea that is connected to your territory (UNCLOS 1982: Art. 3,4).

12 For example, Canada, Denmark and Norway subsequently placed flags in the Arctic as well. Canada has upgraded its Arctic naval and army forces and the US has started measuring the Arctic seabed to substantiate territorial claims (Ryan 2011).

13 There is one central feature of all climate political measures – namely, that they are expert-driven measures (e.g. carbon trading, resilience) to manage a range of risks defined by experts, for example through the 2011 IPCC report *Managing the risks of extreme events and disasters to advance climate change adaptation* (IPCC 2012).

Reference

Anderson, Ben. 2010. "Preemption, precaution, preparedness: anticipatory action and future geographies." *Progress in Human Geography* 34 (6): 777–798.

Ashton, John and Tom Burke. 2004. "The geopolitics of climate change." In *SWP comments*. Berlin: SWP, German Institute for International and Security Affairs.

Baldwin, Andrew. 2014. "The political theologies of climate change-induced migration." *Critical Studies on Security* 2 (2): 210–222.

Barnett, John. 2003. "Security and climate change." *Global Environmental Change* 13 (1): 7–17.

Bettini, Giovanni. 2013. "(In)convenient convergences: 'climate refugees,' apocalyptic discourses and the depoliticization of the debate on climate-induced migration." In *(De-)constructing the greenhouse: interpretive approaches to global climate governance*, edited by Chris P. Methmann, Delf Rothe and Benjamin Stephan, 122–136. London: Routledge.

Bloomfield, Steve. 2007. "Climate change leading to global fight for resources." *The Independent*, June 21, 2007.

Boas, Ingrid. 2014. "Where is the South in security discourse on climate change? An analysis of India." *Critical Studies on Security* 2 (2): 148–161.

Bojanowski, Axel. 2011. "Schlechtes Wetter verursacht Bürgerkriege." *Spiegel Online*, August 24, 2011. Accessed April 30, 2015. www.spiegel.de/wissenschaft/natur/klimaphaenomen-el-nino-schlechtes-wetter-verursacht-buergerkriege-a-782166.html.

Boykoff, Maxwell T., David Frame and Samuel Randalls. 2010. "Discursive stability meets climate instability: A critical exploration of the concept of 'climate stabilization' in contemporary climate policy." *Global Environmental Change* 20 (1): 53–64.

Brassett, James and Nick Vaughan-Williams. 2015. "Security and the performative politics of resilience: critical infrastructure protection and humanitarian emergency preparedness." *Security Dialogue* 46 (1): 32–50.

Bulkeley, Harriet, Johannes Stripple, Roger Hildingsson, Hannes Stephan, Rickard Anderson and Tobias Nielsen. 2011. "Governing the global climate polity: rationality, practice and power: workshop report." Lund, Sweden: Lund University, September 2011.

Burke, Marshall B., Edward Miguel, Shanker Satyanath, John A. Dykema and David B. Lobell. 2009. "Warming increases the risk of civil war in Africa." *Proceedings of the National Academy of Sciences* 106 (49): 20670–20674.

Buzan, Barry and Ole Wæver 2009. "Macrosecuritisation and security constellations: reconsidering scale in securitisation theory." *Review of International Studies* 35 (2): 253–276.

Campbell, Kurt M., Jay Gulledge, J. R. McNeill, John Podesta, Peter Ogden, Leon Fuerth, R. J. Woolsey, Alexander T. Lennon, Julianne Smith and Richard Weitz. 2007. *The age of consequences: the foreign policy and national security implications of global climate change.* Washington DC: Center for Strategic and International Studies.

Carrington, Damian. 2011. "Climate changes double risk of civil war, scientists warn." *The Guardian,* August 25, 2011, 23.

Center for Naval Analysis. 2007. "National security and the threat of climate change." Last Modified 2007. Accessed May 4, 2015. https://www.cna.org/sites/default/files/National Security and the Threat of Climate Change – Print.pdf.

Chandler, David. 2012. "Resilience and human security: the post-interventionist paradigm." *Security Dialogue* 43 (3): 213–229.

Chandler, David. 2014. "Beyond neoliberalism: resilience, the new art of governing complexity." *Resilience* 2 (1): 47–63.

Chelleney, Bahrama. 2008. "Say no to 'NPT' of climate change." *The Japan Times,* August 7, 2008.

Collier, Stephen J. 2009. "Topologies of power Foucault's analysis of political government beyond 'governmentality'." *Theory, Culture and Society* 26 (6): 78–108.

Corry, Olaf. 2014. "From defense to resilience: environmental security beyond neoliberalism." *International Political Sociology* 8 (3): 256–274.

Dalby, Simon. 2013. "The geopolitics of climate change." *Political Geography* 37: 38–47.

Dalby, Simon. 2015. "Climate geopolitics: securing the global economy." *International Politics,* published online, March 20, 2015.

Davenport, Corel. 2014. "Pentagon signals security risks of climate change" *New York Times,* October 13, 2014, A14.

de Goede, Marieke and Sam Randalls. 2009. "Precaution, preemption: arts and technologies of the actionable future." *Environment and planning. D, Society and space* 27 (5): 859.

Dunlap, Alexander and James Fairhead. 2014. "The militarisation and marketisation of nature: an alternative lens to 'climate-conflict'." *Geopolitics* 19 (4): 937–961.

Evans, Brad and Julian Reid. 2013. "Dangerously exposed: the life and death of the resilient subject." *Resilience* 1 (2): 83–98.

Fetzek, Shiloh and Jeffrey Mazo. 2014. "Climate, scarcity and conflict." *Survival* 56 (5): 143–170.

Gladwell, Malcolm. 1996. "The tipping point: why is the city suddenly so much safer – could it be that crime really is an epidemic?" *The New Yorker* 72 (14): 32–38.

Gorbachev, Mikhail and Jean-Michel Severino. 2007. "Climate change raises threat of water wars." *The Japan Times*, June 9, 2007.

Grove, Kevin J. 2010. "Insuring 'our common future?' Dangerous climate change and the biopolitics of environmental security." *Geopolitics* 15 (3): 536–563.

Hajer, Maarten A. 1995. *The politics of environmental discourse: ecological modernization and the policy process.* Oxford: Clarendon Press.

Hartmann, Betsy. 2010. "Rethinking climate refugees and climate conflict: rhetoric, reality and the politics of policy discourse." *Journal of International Development* 22 (2): 233–246.

Hartmann, Betsy. 2013. "Climate chains: neo-Malthusianism, militarism and migration." In *Interpretive approaches to global climate governance: (De)constructing the greenhouse*, edited by Chris P. Methmann, Delf Rothe and Benjamin Stephan, 91–104. London: Routledge.

Hartmann, Betsy. 2014. "Converging on disaster: climate security and the Malthusian anticipatory regime for Africa." *Geopolitics* 19 (4): 757–783.

Heinemann, Pia. 2011. "El Niño fördert blutige Konflikte." *Die Welt*, August 28, 2011, 46.

Hindustan Times. 2009. "Climate change a more serious threat than Islamic radicalism." October 2, 2009.

Howarth, David and Steven Griggs. 2008. "Metaphor, catachresis and equivalence: The rhetoric of freedom to fly in the struggle over aviation policy in the United Kingdom." In *Political language and metaphor*, edited by Terell Carver and Jernej Pikalo, 197–211. London: Routledge.

Howden, Daniel. 2007. "Diplomats warned that climate change is security issue, not a green dilemma." *The Independent*, December 6, 2007, 36.

Hulme, Mike. 2008. "The conquering of climate: discourses of fear and their dissolution." *The Geographical Journal* 174 (1): 5–16.

Ide, Tobias and Jürgen Scheffran. 2014. "On climate, conflict and cumulation: suggestions for integrative cumulation of knowledge in the research on climate change and violent conflict." *Global Change, Peace and Security* 26 (3): 263–279.

IPCC, Intergovernmental Panel on Climate Change. 2012. *Managing the risks of extreme events and disasters to advance climate change adaptation. A special report of working groups I and II of the intergovernmental panel on climate change.* Cambridge: CUP.

IPCC, Intergovernmental Panel on Climate Change. 2014. *Climate change 2014: impacts, adaptation, and vulnerability. Working group II contribution to the fifth assessment report of the intergovernmental panel on climate change.* New York: Cambridge University Press.

Japan Times, The. 2009. "Arctic heating up." October 4, 2009. Accessed April 30, 2015. www.japantimes.co.jp/opinion/2009/10/04/editorials/arctic-heating-up/.

Johnstone, Sarah and Jeffrey Mazo. 2011. "Global warming and the Arab Spring." *Survival* 53 (2): 11–17.

Jordan, Mary. 2007. "Former vice president depicts climate change as 'threat to the survival of our civilization'." *Washington Post*, December 10, 2007, 14.

Joseph, Jonathan. 2013. "Resilience as embedded neoliberalism: a governmentality approach." *Resilience* 1 (1): 38–52.

Knickerbocker, Brad. 2007. "Could warming cause war?" *Christian Science Monitor*, April 19, 2007, 2.

Lacy, Mark. 2005. *Security and climate change: international relations and the limits of realism*. New York: Routledge.

Lenton, Timothy M. 2011. "Beyond 2 C: redefining dangerous climate change for physical systems." *Wiley Interdisciplinary Reviews: Climate Change* 2 (3): 451–461.

Lenton, Timothy M., Hermann Held, Elmar Kriegler, Jim W. Hall, Wolfgang Lucht, Stefan Rahmstorf and Hans Joachim Schellnhuber. 2008. "Tipping elements in the Earth's climate system." *Proceedings of the National Academy of Sciences* 105 (6): 1786–1793.

Liverman, Diana M. 2009. "Conventions of climate change: constructions of danger and the dispossession of the atmosphere." *Journal of Historical Geography* 35 (2): 279–296.

Lohmann, Larry. 2005. "Malthusianism and the terror of scarcity." In *Making threats: biofears and environmental anxieties*, edited by Betsy Hartmann, Banu Subramaniam and Charles Zerner, 81–98. Oxford: Rowman and Littlefield.

Maas, Achim and Dennis Tänzler. 2009. *Regional security implications of climate change. A synopsis*. Berlin: Adelphi research.

Mabey, Nick. 2008. *Delivering climate security: international security responses to a climate changed world. Edited summary*. London: E3G.

Mabey, Nick. 2011. *Degrees of risk: defining a risk management framework for climate security*. London: Third Generation Environmentalism Limited.

MacFarquhar, Neil. 2011. "No consensus on addressing climate shift." *New York Times*, July 21, 2011, 9.

Mason, Michael. 2014. "Climate insecurity in (post) conflict areas: the biopolitics of United Nations vulnerability assessments." *Geopolitics* 19 (4): 806–828.

Mert, Aysem. 2013. "Discursive interplay and co-constitution: carbonification of environmental discourses." In *(de-)constructing the greenhouse: interpretive approaches to global climate governance*, edited by Chris P. Methmann, Delf Rothe and Benjamin Stephan, 23–42. London: Routledge.

Methmann, Chris and Angela Oels. 2015. "From 'fearing' to 'empowering' climate refugees: governing climate-induced migration in the name of resilience." *Security Dialogue* 46 (1): 51–68.

Methmann, Chris P. and Delf Rothe. 2012. "Politics for the day after tomorrow: the political effect of apocalyptic imaginaries in global climate governance." *Security Dialogue* 43 (3): 323–344.

Methmann, Chris P. and Delf Rothe. 2013. "Apocalypse now! From exceptional rhetoric to risk management in global climate politics." In *Interpretive approaches to global climate governance: (de-)constructing the greenhouse*, edited by Chris P. Methmann, Delf Rothe and Benjamin Stephan, 105–121. London: Routledge.

Nayak, Malathi. 2011. "Threat of climate-change disasters looms large for Bangladesh." *Washington Post*, February 6, 2011, 5.

Newell, Peter. 2006. *Climate for change: non-state actors and the global politics of the greenhouse*. Cambridge: CUP.

Nonhoff, Martin. 2007. "Politische Diskursanalyse als Hegemonieanalyse." In *Diskurs – Hegemonie – radikale Demokratie: Zum politischen Denken von Ernesto Laclau und Chantal Mouffe*, edited by Martin Nonhoff, 173–193. Bielefeld: Transkript.

Oels, Angela. 2013. "Rendering climate change governable by risk: from probability to contingency." *Geoforum* 45: 17–29.

Randalls, Samuel. 2010. "History of the 2 C climate target." *Wiley Interdisciplinary Reviews: Climate Change* 1 (4): 598–605.
Reid, Julian. 2012. "The disastrous and politically debased subject of resilience." *Development Dialogue* 58: 67–80.
Reuveny, Rafael. 2007. "Climate change-induced migration and violent conflict." *Political Geography* 26 (6): 656–673.
Rodgers, Walter. 2010. "War over the Arctic? Global warming skeptics distract us from security risks." *Christian Science Monitor*, February 3, 2010.
Rothe, Delf. 2012. "Security as a weapon: how cataclysm discourses frame international climate negotiations." In *Climate change, human security and violent conflict*, edited by Jürgen Scheffran, Michael Brzoska, Hans Günter Brauch, Peter M. Link and Janpeter Schilling, 243–258. Berlin/Heidelberg: Springer.
Ryan, Jacquelyn. 2011. "As Arctic melts, US ill equipped to tap resources." *Washington Post*, January 10, 2011, 2.
Sappenfield, Mark. 2007. "Climate may heat conflict, too." *Christian Science Monitor*, December 6, 2007, 13–17.
Scheffran, Jürgen, Michael Brzoska, Jasmin Kominek, P. Michael Link and Janpeter Schilling. 2012. "Climate change and violent conflict." *Science* 336 (6083): 869–871.
Schmidt, Jessica. 2014. "Intuitively neoliberal? Towards a critical understanding of resilience governance." *European Journal of International Relations* 21 (2): 402–426.
Schwartz, Peter and Doug Randall. 2003. "An abrupt climate change scenario and its implications for United States national security." Washington, DC: Global Business Network.
Sethi, Nitin. 2011. "India prevents UNSC from hijacking climate talks." *The Times of India*, July 23, 2011. Accessed April 30, 2015. http://timesofindia.indiatimes.com/home/environment/developmental-issues/India-prevents-UNSC-from-hijacking-climate-talks/articleshow/9331057.cms.
Solana, Javier. 2008. *Climate change and international security: paper from the High Representative and the European Commission to the European Council.* Brussels: European Commission.
Spiegel Online. 2006. "Klimakonferenz: Annan verlangt mehr politischen Mut im Kampf gegen Klimawandel." Last Modified November 15, 2006 Accessed April 30, 2015. www.spiegel.de/wissenschaft/natur/klimakonferenz-annan-verlangt-mehr-politischen-mut-im-kampf-gegen-klimawandel-a-448577.html.
Spiegel Online. 2009. "Klimawandel: Forscher warnen vor beispiellosen Flüchtlingswellen." Last Modified June 10, 2009 Accessed April 30, 2015. www.spiegel.de/wissenschaft/natur/klimawandel-forscher-warnen-vor-beispiellosen-fluechtlingswellen- a-629680.html.
Spranz, Nina M. 2007. "'Es war eine Überraschung': Wofür der Weltklimarat den Friedensnobelpreis bekam." *Die Welt*, October 13, 2007.
Stripple, Johannes. 2005. "Climate change after the international: rethinking security, territory and authority." Lund, Sweden: Department of Political Science, University of Lund.
Süddeutsche Zeitung Online. 2010. "Klimakonferenz in Kopenhagen: Die Menschheit vor dem Ernstfall." Last Modified May 17, 2010 Accessed April 30, 2015. www.sueddeutsche.de/wissen/klimakonferenz-in-kopenhagen-die-menschheit-vor-dem-ernstfall-1.147886.

Swyngedouw, E. 2010a. "Apocalypse forever? Post-political populism and the spectre of climate change." *Theory Culture and Society* 27 (2–3): 213–232.

Swyngedouw, Erik. 2010b. "Trouble with nature: 'ecology as the new opium for the masses'." In *The Ashgate research companion to planning theory: conceptual challenges for spatial planning*, edited by Jean Hillier and Patsy Healey, 299–318. Farnham: Ashgate.

Taleb, Nassim Nicholas. 2010. *The black swan: the impact of the highly improbable fragility*. New York: Random House.

Tenbrock, Christian. 2009. "Klimaflüchtlinge: Der große Treck." *Die Zeit*, October 8, 2009.

Theisen, O. M., H. Holtermann and H. Buhaug. 2012. "Climate wars? Assessing the claim that drought breeds conflict." *International Security* 36 (3): 79–106.

UK Mission, United Kingdom Mission at the United Nations. 2007. "Energy, security and climate: security council open debate: UK concept paper." Accessed May 4, 2015. https://unfccc.int/files/application/pdf/ukpaper_securitycouncil.pdf.

UNCLOS. 1982. *United Nations convention on the law of the sea*. New York: United Nations.

UNFCCC. 2010a. *Report of the conference of the parties on its sixteenth session, held in Cancun from 29 November to 10 December 2010*. Accessed May 5, 2015. http://unfccc.int/resource/docs/2010/cop16/eng/07a02.pdf.

UNFCCC, United Nations Framework Convention on Climate Change. 2010b. *Report of the conference of the parties on its sixteenth session, held in Cancun from 29 November to 10 December 2010*. FCCC/CP/2010/7/Add.1.

UNFCCC, United Nations Framework Convention on Climate Change. 2012. *Report of the adaptation committee of work*. FCCC/SB/2012/3.

UNGA, United Nations General Assembly. 2009a. *Climate change and its possible security implications: Report of the Secretary-General*. New York: United Nations General Assembly.

UNGA, United Nations General Assembly. 2009b. Sixty-third session, 85th plenary meeting Wednesday, June 3, 2009.

United Nations. 2014. *Press release: climate summit scales up adaptation and resilience support for most vulnerable countries*. New York: United Nations.

UNSC, United Nations Security Council. 2007a. Sixty-second year, 5663rd meeting, Tuesday, April 17, 2007.

UNSC, United Nations Security Council. 2007b. Sixty-second year, 5663rd meeting, Tuesday, April 17, 2007, Resumption.

UNSC, United Nations Security Council. 2011a. Sixty-sixth year, 6587th meeting, Wednesday, July 20, 2011.

UNSC, United Nations Security Council. 2011b. Sixty-sixth year, 6587th meeting, Wednesday, July 20, 2011, Resumption.

Walker, Jamie and Andrew Faulkner. 2007. "Keelty warning on global warming." *The Australian*, September 25, 2007, 4.

Walker, Jeremy and Melinda Cooper. 2011. "Genealogies of resilience from systems ecology to the political economy of crisis adaptation." *Security Dialogue* 42 (2): 143–160.

WBGU, German Advisory Council on Global Change. 2009. *Climate change as a security risk*. London: Earthscan.

6 The Euro-Mediterranean region

Not in my backyard

> Since the EU's neighbours include some of the most vulnerable regions to climate change, e.g. North Africa and the Middle East, migratory pressure at the European borders and political instability and conflicts could increase in the future.
>
> (Solana 2008: 6)

In this chapter the focus shifts from the international climate security discourse to the discourse on the climate impacts in the MENA (Middle East and North Africa) region.[1] The climate security discourse on the Mediterranean region[2] represents a paradigmatic case for the analysis at hand. First of all, the southern Mediterranean is regarded as one of the regions most threatened by negative impacts of climate change such as droughts, floods or agricultural decline (Phone interview with a Geneva-based environmental security expert, September 15, 2011). At the same time "the Mediterranean is like a pilot region for a lot of regions in the world" (Phone interview with an Abu Dhabi-based renewable energy analyst, November 10, 2011). Furthermore, the quote beginning this chapter already indicates why the region is particularly important: namely, due to its geographic proximity to the European Union, which is one of the forerunners in the discourse (Phone interview with a Bonn-based climate security analyst, September 19, 2011). In other words, it is the perception of the region as the EU's *southern flank* and a potential gateway for security risks that makes it a particularly prominent case in the discourse.

The latter assumption is supported by the fact that actors in the southern Mediterranean itself, i.e. in the countries most threatened, are seldom concerned about possible climate related challenges (Interview with a member of the German Foreign Office, Berlin, September 14, 2011). In this chapter I show that the climate change and security discourse on the Mediterranean is an *export-object* as part of the EU's Mediterranean politics. I therefore begin by outlining the governmental landscape of EU's Mediterranean politics and then show how climate security is imbedded within this political field. The second part of the chapter then outlines two logics of securitization:

first, *climate change and security as an export* (from the EU to the MENA region) and second, *climate change and security as a translation.*

Competing security discourses in the EU's Mediterranean politics[3]

Just as with any other political space, the Mediterranean is not marked by any essential common features: "There is no primeval Mediterranean unity, neither historical, nor in cultural or ecological terms" (Öktem 2010: 31). On the contrary, Mediterranean countries are marked by a great diversity economically and politically as well as culturally, which has, in the past, not uncommonly led to a climate of distrust and political tensions. Past European approaches to *Mediterranean regional politics* have been interpreted as political projects to modify this situation of fragmentation, which, due to the proximity of the southern Mediterranean, was seen as a security problem. The EU's regional politics in the Mediterranean have been shaped by two competing security discourses (see Malmvig 2004). First, the Euro-Mediterranean Partnership (EMP) initiated by the so-called Barcelona convention in 1995 builds upon a discourse of *cooperative security*, which evolved in Europe following the success of the OSCE (Adler and Crawford 2004). The overall rationale of the EMP was to establish a "Euro-Mediterranean zone of peace and stability" (European Commission 1994) through functional cooperation in three different "baskets" on security, economic and civil society cooperation (Jünemann 2009: 41–48).

The EMP can be read as a European attempt to repeat the success story of the OSCE by establishing a security community in the Mediterranean region. With this perspective, region building represents an endeavor to establish novel collective identities through cooperative diplomacy, 'talk-shops' and similar measures (Adler and Crawford 2004: 17). This discourse of cooperative security constructs a range of global security challenges that threaten every country equally in the region and that can only be addressed collectively. Such common challenges are, for example, Islamist terrorism, international migration, organized crime, and, as we will see below, climate change. The threatened subject in this security discourse is made up of "two equitable 'communities' [EU and MENA] joined by the common referent of the Mediterranean" (Malmvig 2004: 11). The Union for the Mediterranean (UfM), founded in 2008, was informed by the same discourse of cooperative security. The UfM initiated by then French Premier Nicolas Sarkozy comprises a variety of different policy fields, such as energy, environment, education and economic policy. Yet the UfM does not seek to address them in a holistic manner, but rather concentrates on a few lighthouse projects, thereby circumventing problematic political topics and hoping to spur further cooperation. The hope was that such rather apolitical, technocratic cooperation projects could spur regional integration and at the same time bypass complicated political matters – such as migration or the Israel–Palestine conflict.

A second, opposing security discourse, which could be called a *liberal modernization discourse*, has also informed the regional politics of the EU. It basically draws on an antagonism constructed between the European homeland and its unstable southern neighborhood (Pace 2004: 293). Following this framing, threats such as terrorism, radical Islam or migration do not represent a common challenge for both shores of the Mediterranean but rather arise from the economic deficits and social grievances that mark MENA countries (Malmvig 2004: 14). The *otherness*, or economic and political *inferiority*, of the southern Mediterranean here becomes a source of insecurity for the EU. This 'othering' is not only important as it constitutes the EU's own identity as a community of peace and stability but also affects the political practices of the EU in the region (Diez 2005: 628).

A liberal modernization discourse has clearly informed the extension of the European Neighborhood Policy (ENP) to the EU's southern environment, as a complement to the EMP in 2004. The overall goal of the ENP was a) to increase European security by a "ring of well-governed countries" (EU Security Strategy 2003: 8) on the outskirts of the EU's external borders (Jeandesboz 2007: 399–400), and b) to "prevent the emergence of new dividing lines, through a variety of means including more cross-border cooperation" (Smith 2005: 763). With increasing internal integration and the abolishment of internal borders, moreover, safeguarding the external frontiers of the EU became more important (Huysmans 2000: 758–762). On the basis of bilateral action plans, the ENP offers neighboring countries positive incentives such as access to the common market or increased mobility to facilitate good governance. The central rationale was to establish a privileged space around its external borders that allows for the free circulation of economic goods, investments and services (Dimitrovova 2008: 3). According to the rationality of liberal modernization, such liberalization measures would foster political democratization (see Horst 2013). In sum, "this EU narrative constructs a relational triad between economic prosperity, stability and peace" (Pace 2009: 43).

Through the events of the Arab Spring in early 2011, all initiatives of the EU's Mediterranean policy became discredited, given the long cooperation of the EU with authoritarian governments in MENA countries (see Jünemann *et al.* 2013). The Arabellions and the shockwaves they produced posed a serious challenge to both the assumption of the existence of a common Mediterranean region as well as to the belief in liberal modernization. The EU reacted promptly and adopted two proposals – for a "A partnership for democracy and shared prosperity with the Southern Mediterranean" (European Commission 2011b) and a "A new response to a changing Neighbourhood" (European Commission 2011a) that should bring the promotion of democracy and support for civil society back to the top of the agenda. In particular these proposals introduced the 'more for more approach' – a mechanism of positive conditionality – promising incentives for cooperating states in terms of increased financial assistance for the development of civil societies, increased mobility as well as better access to the EU single market. However, as different scholars have shown, real change has been marginal not only due to path dependencies and structural obstacles to change (Bauer 2015: 27) but also because the blunt realities in countries such as Libya, Syria and to a lesser degree Egypt soon shifted the EU's attention back to the issue of security and stability.

The governmental landscape of region building in the Mediterranean

When it comes to the plurality of political practices and technologies that have informed Euro-Mediterranean politics, I would broadly distinguish between two major strands of governmental practices, which in reality sometimes overlap. On the one hand we could identify different forms of advanced liberal government and on the other hand practices of biopolitical management, oscillating between rather liberal and more repressive forms of control. Practices of advanced liberal government, drawing on a government at a distance, are probably the most visible ones in EU's Mediterranean policy. First, this is meant in a very empirical sense: the cornerstone of the Association Agreements adopted together with MENA countries in the early 2000s was the abolition of trade barriers (Jünemann 2009: 44). Besides the EU Association Agreements, which set the legislative frame for such a free-trade zone, the zone was to be established by a series of technocratic or regulatory cooperative instruments in different sectors (Bicchi 2006: 149). In the energy sector, for example, there were initiatives to achieve a market opening and liberalization in MENA countries but also measures to achieve a technical and regulatory harmonization between MENA countries and the EU members.

From a governmentality perspective, however, neoliberalism is more than de-regulation and laissez-faire politics. A governmentality perspective instead sheds light on the highly productive character of neoliberal practices (Traub-Werner 2007: 1443). Advanced liberal practices are productive as they construct subject positions of self-responsible, entrepreneurial and innovative agents within privileged spaces (Kurki 2011: 349). This becomes most obvious in the European practices that seek to empower civil societies in southern Mediterranean neighborhood countries. This implies not an approach of just 'letting it be' but actively bringing about the *right* civil society (Kurki 2011: 362), i.e. a civil society modeled after the European prototype.

At the same time we can identify practices and technologies in the EU's politics of the region that take the Mediterranean populations (or parts of it) and other life-related systems such as energy supply or the environment as a governmental subject. As I outlined in the previous section, economic and political underperformance and the resulting social grievances were problematized as a source of danger for the EU. As a consequence, the EU became concerned with phenomena such as economic performance, population growth, the spread of disease, etc. Euro-Mediterranean politics, then, sought to establish political institutions to access and manage these parameters. A case in point is the development of a common civil protection infrastructure within the EMP and the UfM (Bremberg 2010b: 175–178). The boundaries to advanced liberal forms of government, however, are fluid. For example, in many respects the state – or the EU – is perceived as being either unwilling or incapable of optimizing the well-being of the population. In these cases it is believed that the market could do better. This is precisely the rationale behind the ENP, which seeks to establish a privileged space for the circulation of goods, capital, services and people (see Dimitrovova 2008: 53–54).

That being said, the creation of free circulation also produces a range of dangers. While a privileged space of free circulation is a political solution to some problems, it also produces others. It is here that laissez-faire forms of advanced liberal government reach their limit. As Dimitrovova (2008) notes, the ENP does not just produce a privileged space of free movement but clearly distinguishes between good (capitals, goods, services) and bad (people, migrants) forms of circulation. While the former is promoted, the latter has to be identified and prevented. In this case biopolitical control and management take on a rather repressive form of power. Through 'law enforcement cooperation' on migration, e.g. through the Action Plans of the ENP, the EU has externalized "immigration control beyond their borders towards southern Mediterranean countries, which are thus being co-opted into the role of 'gatekeepers' or 'buffer states'" (Lutterbeck 2006: 73). In fact, this is meant to externalize repressive forms of migration control, e.g. through the establishment of detention camps in the territories of authoritarian North African states (see White 2011: 114). Biopolitical management thus introduced a series of clear-cut antagonisms (inside/outside; privileged/unprivileged; legal/illegal, etc.), bringing in a logic of antagonistic security.[4]

The region in climate security discourse

So far I have shown that the politics of the region in the Mediterranean was informed by different security discourses that came along with competing representations of the Euro-Mediterranean space: a common Mediterranean; MENA as EU's dangerous neighborhood; and a buffer to an even more dangerous outside. One of the main reasons why the discourse on climate change and security on the region gained such momentum in Europe is that it ties in with these different security discourses and spatial frames.

One or many Mediterraneans? Climate security and space

First of all, one can clearly see that climate change and security discourse storylines resonate with a cooperative security discourse. This becomes apparent in articulations following a logic of equivalence, constructing a similarity of countries around the Mediterranean with respect to their vulnerability to major environmental problems (Phone interview with a member of the European External Action Service, November 25, 2011). Environmental resemblances have informed the construction of a Mediterranean region from the very beginning (Brauch 2010: 487). Now, from a discourse theoretical perspective it is clear that any identity could not be established solely in positive terms. A regional identity for the Mediterranean draws on a process of *othering* – a differentiation from other political entities (Korkmaz 2008: 142). The climate change and security discourse presents dangerous climate change as a personified enemy of the whole region. In the face of a "survival dilemma that transgress[es] national borders and environmental boundaries," the Mediterranean is being constructed as a "community of fate" (Brauch 2010: 20). Climate change is put alongside a series of "common Euro-Mediterranean" challenges (Malmvig 2004: 9) such as terrorism, radicalism and illegal immigration.

This equivalence framing, however, is not the only spatial representation in the discourse and far from being uncontested (Phone interview with an Amsterdam-based climate and energy activist, September 26, 2011). The reference to *the Mediterranean* or the MENA region as a climate change or climate security hotspot is almost entirely restricted to European articulations. This corresponds to the assumption that the Mediterranean region is a European invention – articulated as a counter-model to the established concept of the Middle East region, which is promoted and favored by the US (Malmvig 2004: 10). Many other sources in climate change and security discourse, on the other hand, articulate rather differential spatial concepts of the same geographic area. In these articulations it is not the Mediterranean region as a whole that is threatened by climate change. Rather, they problematize the security implications of climate change in the Middle East (Brown and Crawford 2009; Center for Naval Analysis 2007), North Africa (NIC 2009), MENA (Maas and Tänzler 2009: 4–6) or sometimes 'the Arab Region' (Elasha 2010).[5] In short, there is an ongoing discursive struggle over different spatial representations of climate security implications in the geographical area around the Mediterranean Sea.

Furthermore, even articulations that take the Mediterranean as a common referent point always draw on a second, differential construction of the region as expressed in these examples: "Should climate change spark conflicts one of the most likely battlegrounds could be *the line dividing* the Mediterranean" (MedSec 2009: 10, emphasis added); and "The Mediterranean is a fault line of political, economic, societal and environmental insecurity" (Brauch 2010: 49). In this differential framing, dangerous climate change is not the external threat that unites the region, but a process that sets the southern and northern shores of the Mediterranean apart. This spatial framing is differential as it clearly distinguishes between (southern) Europe and the southern part of the Mediterranean by drawing on a series of binary oppositions:

> In southern Europe, environmental changes such as droughts and heatwaves are unlikely to lead to outbreaks of violence in the foreseeable future. The North African countries, by contrast, are far more vulnerable, because their environmental situation is significantly worse than that of southern Europe and their societies have less problem-solving capacity.
>
> (WBGU 2009: 136)

From a discourse theoretical perspective of asking what is being problematized, we clearly see that it is the climate change impacts in the southern Mediterranean whose "situation can have consequences for the stability of the entire region" (WBGU 2009: 136). The southern Mediterranean becomes the primary source of anxiety for Europe. The reason for this anxiety is the degree of proximity, interrelatedness and complexity that is associated with the concept of a region. Furthermore, "the hazards of climate change will exacerbate the unequal share-out [...] between the northern and southern shores of the Mediterranean" (MedSec 2009: 11).

In a more radical version the southern Mediterranean even becomes a (source of) threat to Europe itself: "Europe's main interest lies in preventing spillover from climate change-induced instability from North Africa, principally in terms of migration, but also access to energy, terrorism and transnational crime" (NIC 2009: 28). Here the equivalence chain is completely broken up; the two sides of the Mediterranean Sea are not considered to constitute a single region anymore. Nevertheless, even in such an antagonistic framing, the southern Mediterranean is not constructed as a political enemy but rather as a *special interest area*. This is because it represents Europe's 'southern flank' and a 'gateway' for a series of structural dangers like migration: "migration from other regions through Northern Africa to reach Europe (transit migration) is likely to intensify" (Solana 2008: 6). European Mediterranean politics hence aims to transform the Mediterranean from a channel into a buffer zone: "Europe seeks to build a *cordon sanitaire*, against Sub-Saharan migration in North Africa" (NIC 2009: 31).

Collective symbols play a particular role in constructing different spatial framings of the climate security implications in the Mediterranean. Of particular importance are container, boat or body metaphors that help in constructing political communities as objects. The different spatial framings that could be identified in the climate change and security discourse are summarized in Figure 6.1.

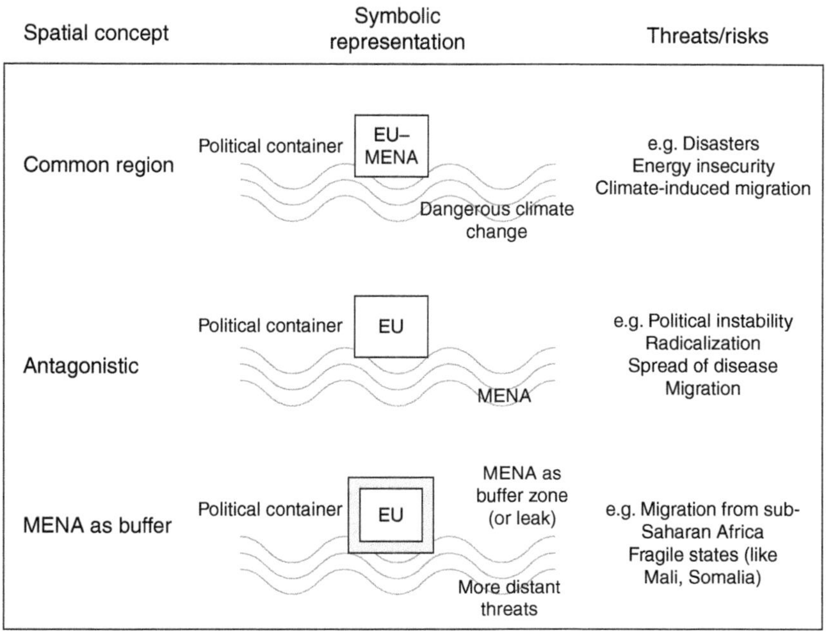

Figure 6.1 Spatial representations of the Mediterranean in climate security discourse (source: author).

Mapping the Mediterranean into existence

The spatiality of climate change and security implications is not constructed solely through linguistic articulations. Different practices play an important role in constructing this spatiality. A practice that is of particular importance in the climate change and security discourse is mapping or cartography (Crampton 2001).[6] The practice of mapping can be best explained using the case of the MedSec (Mediterranean Security) initiative (MedSec 2009). On June 25–27, 2009 this network of non-governmental and international organizations, think tanks, the OSCE and research institutions (MedSec 2009: 48–49) conducted a workshop *Participatory Assessment of Environment and Security Issues in the Southern Mediterranean Region* in Barcelona. The workshop brought together experts and bureaucrats from different institutional contexts and from different countries from Europe and the MENA countries. The centerpiece of the workshop was a mapping exercise:

> to visually highlight the factors and issues based on input from experts from the region [...]. The purpose of the maps is to reveal and communicate priorities and inter-linkages in a format that is easily understandable, lending itself to swift analysis.
>
> (MedSec 2009: 7)

In this short quote some crucial points are already raised. A map is a set of signs or discursive device that, through its direct visual accessibility, presents very complex narratives in a very accessible way (Wood and Fels 1986). Furthermore, maps allow the use of basal icons and markers that are cognitively linked with certain affective states – for example the color red or the blizzard icon usually symbolizes danger. The discursive effect of maps is to present discursive storylines in a materialized or objectified form: first, because they transform storylines in a very literal sense into objects – maps that can be viewed and touched; second, especially in an age of satellite navigation, because maps are usually taken as an objective representation of reality (Crampton 2001: 698; Harley 1989: 10).

The Barcelona Experts' workshop further advanced this logic by conducting a collective mapping exercise. In this workshop participants were split into smaller groups, given blank or topographic maps of the Mediterranean, and asked to "name and locate as precisely as possible what they see as major environment and security issues" (MedSec 2009: 39). Here, it is not only that, by looking at a map, participants adopted certain storylines about future climate security impacts in the region, but that they themselves became part of the storytelling. By interacting with the map and the drawings of others participants, they likewise internalized, reproduced and changed the articulations made by others, such as the expert knowledge they were given as an input. The aim was to raise awareness and to activate the self-potential of the participants from the Mediterranean region (MedSec 2009: 7).

Figure 6.2 is an example of one of the results of the mapping exercise. Most obviously, these maps *construct a specific spatial representation* of the climate change and security discourse in the Mediterranean. In the given example, the focus clearly is on the southern Mediterranean – Europe at the upper front of the map is only represented as a receiver of resources, energy and people. What can be further seen from the picture is that it is not a static representation of a geographic region but that it accounts for movement and projected change through arrows and other icons (one arrow represents migration pressures, another the spread of desertification). The original used colors as markers for certain affective judgments: red and orange for the negative trends, green for the fight against desertification. The map does not just depict geographic and political facts; it tells whole narratives in a simple manner. For example, it anticipates a future security risk by placing an icon for a projected nuclear plant close to cities such as Rabat or Algiers that will come "under high population pressure" (MedSec 2009: 41). It almost unnoticeably fuses the present – such as border disputes between Morocco and Algeria – with an imagined future – such as the spread of desertification.

Maps as seemingly objectified representations of climate security storylines also play an important role in authorized articulations in the climate security discourse. Figure 6.3, from the summary for policymakers of the

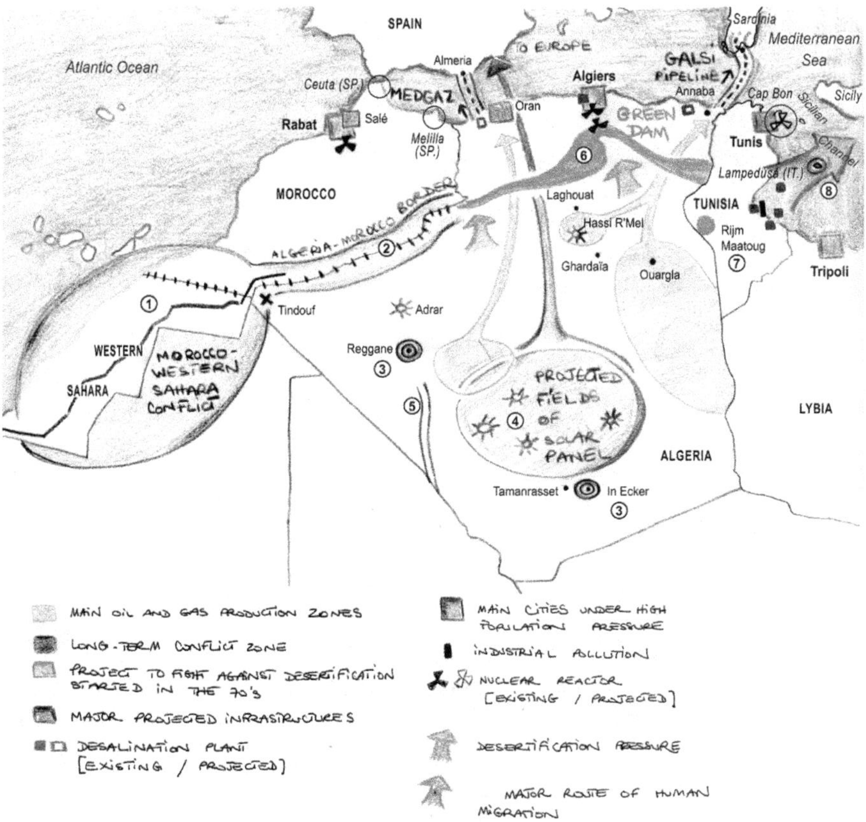

Figure 6.2 Outcomes of the participatory mapping exercise (reprinted with kind permission of zoï environment, copyright 2009).

2008 German WBGU report, is an illustrative example in this respect. The world map identifies different future climate change hotspot regions with color-coded markers. As we can see, the whole space surrounding the Mediterranean Sea is marked as being under high risk of future droughts. And this is not a triviality as such maps ultimately tell decision makers where the areas of interest will lie in the future.

A good example of the spatial construction of climate change risk zones is the so-called 'overlap approach,' which basically consists of the act of laying different maps on top of each other. For example, you produce a map, which marks the projected climate impact hotspots in a certain space, just like Figure 6.3 does. Upon this map, then, you place another (transparent) map, which marks socio-political developments. The product of this, following articulations in the climate security discourse, is a comprehensive, more elaborate and more directly accessible picture of

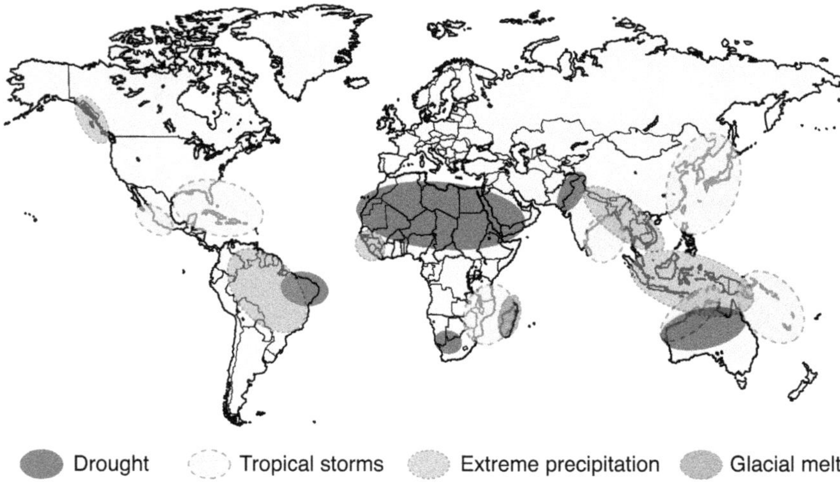

Figure 6.3 Future climate hotspots in different world regions (source: WBGU 2009, p. 161. Reprinted with kind permission of Taylor & Francis and the WBGU, copyright 2009).

future spatialities of climate security implications. This is perfectly expressed in the following extract from an article reporting from a high-level workshop on climate security in the UK:

> An internal presentation to senior diplomats at the Foreign Office listed every recent, serious breakdown of civil order around the world and mapped it against those countries hardest hit by climate change. The fit was almost perfect. One of the diplomats present said there was an "audible intake of breath" from the audience when the slide was shown.
>
> (Howden 2007: 36)

Figure 6.4 is an example of the result of the overlap approach in the (southern) Mediterranean. Compared to Figure 6.3, the inclusion of socio-economic factors shifts the focus away from the Mediterranean towards North Africa. The latter now is represented as a climate security hotspot (symbolized by the dark gray shape) characterized by the threefold threat of climate migration, water scarcity and food insecurity. The assumption in producing a more elaborate picture through the overlap approach is, nevertheless, questionable. What the approach mainly does is to reify and objectify the artificial distinction between the natural and the social world (see Latour 1993). It thus represents a good example of how nature–society relations are imagined. On the one hand a picture of natural impacts that completely abstracts from any socio-economic background is painted. And on

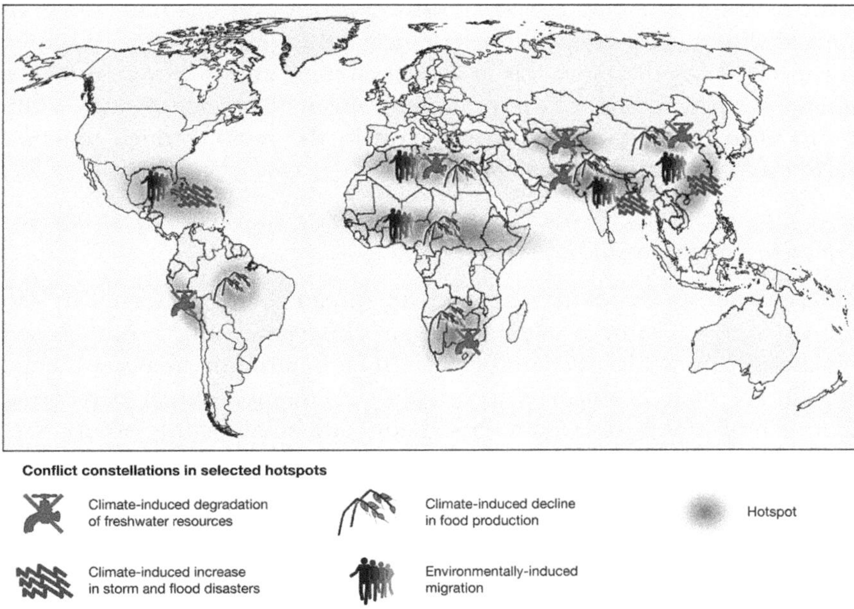

Figure 6.4 Conflict constellations in selected hotspots (source: WBGU 2009, p. 163. Reprinted with kind permission of Taylor & Francis and the WBGU, copyright 2009).

the other hand, these natural changes, in a second step, meet with the socio-economic conditions on the ground.

Securitizing logic I: climate change and security as an export

Due to the outlined spatial framing of climate security discourses that match with the discourses in EU's politics of the region, policy measures to deal with the threat of climate change in the Mediterranean depart from the exceptional measures of the Copenhagen School. It is not that the EU shuts itself off from the southern Mediterranean and does not seek military solutions to the anticipated problems. Exactly the opposite is the case: the EU tries to integrate MENA countries in its approach to managing climate-induced migration and energy insecurity, as we see below. The problem for the EU with such a cooperative security approach is, however, that MENA governments must be willing to cooperate on these issues. There must be a discourse coalition sharing the storylines of climate change as a security threat at least in larger parts of the political community of the Mediterranean. However, in Chapter 4 I showed that climate change and security rationales thus far have not produced much resonance in southern Mediterranean countries. Hence, environmental security concerns have not been addressed and discussed in previous frameworks of Euro-Mediterranean cooperation (Brauch 2012: 749–750). The crux for European actors therefore was to raise awareness and win over voices for climate security in the southern Mediterranean. This section studies different means of discursive dissemination in the region, which together make up a *securitization logic of an export*.

Exporting security storylines

Different European actors have tried to spread the climate security discourse in the southern Mediterranean (Interview with a Bonn-based climate security analyst, September 19, 2011). In terms of discourse theory, the aim of such practices is to establish a common threat perception (Phone interview with a Geneva-based environmental security expert, September 15, 2011), a political identity and will by expanding a discourse coalition on climate security across Europe's southern borders. Practically speaking, this works for example through regional expert workshops. The experts from different ministries or civil society organizations from the MENA region should internalize the storylines discussed at such events and function as *knowledge brokers* who spread the discourse in their home country. A case in point is the Mediterranean Security Partnership with its participatory assessment of environment and security issues in the southern Mediterranean introduced above.[7] The workshop, with regional experts from ministries of MENA countries, aimed to raise the awareness of southern Mediterranean actors for the climate security consequences in their countries without prescribing anything top down:

> A key factor will be the ability to mainstream knowledge rather than preaching to the converted, [...] an assessment of such calibre will only be possible once the countries of the region acknowledge the need for better understanding and greater awareness.
>
> (MedSec 2009: 7)

Exporting climate security to the Mediterranean is neither a designated EU-driven approach nor a coherent program but rather a logic that marks the practices of different governmental and non-governmental actors. Nevertheless, the EU as an actor plays an important role in this overall pattern. At the EU level, regional proposals and initiatives are embedded within a broader 'climate change and international security' (CCIS) process, initiated by the EC in 2008 (Solana 2008). A key element of CCIS for the EU is to work with other regional organizations and to raise awareness about potential security threats arising in vulnerable regions (Solana 2008: 11). The approach of the EU could be understood as a mainstreaming of climate security: "CCIS [climate change and international security] has become a standard item on the agenda of third party meetings in the Mediterranean and Middle East Region at all levels and in a variety of bi- and multilateral formats" (Council 2009: 9). The governmental rationality herein perfectly matches the cooperative security discourse outlined above. In the words of the Commission:

> the EU successfully initiated bilateral and regional dialogue on CCIS. It also succeeded in disseminating information, developing a common understanding and creating platforms for interaction with multiple stakeholders concerning the main threats of climate change. This also highlighted the role of the EU as a global leader on CCIS and its added value as partner and facilitator.
>
> (Council 2009: 5)

We can conclude that some parts of a *European discourse coalition on climate security* actively sought to win over further voices from the southern Mediterranean in order to create a joint threat perception.

Common practices and functional cooperation

Besides this export of climate security storylines, there is a second transfer mechanism, which rather refers to the *export of practices* in a *depoliticized* and *technical* way. Functional cooperation between specialized experts at an administrative level, for example on policing migration, renewable energy cooperation or civil protection, did not necessarily require a simultaneous export of climate security storylines but functions even if there is no common language or understanding of climate security (Phone interview with a Geneva-based environmental security expert, September 15, 2011). This cooperation at the administrative level represents a form of depoliticization in the perspective developed in this book (Chapter 3). On the one hand, it circumvents political contestation and antagonism (the Schmittian element of the political) by bypassing political discourse and focusing on consensual issues. On the other hand, it also suppresses the political in Arendtian terms based on open discourse and the constitution of shared problem perspectives and a common political will.

A good example is cooperation on civil protection. For the EU such cooperation clearly contributes to the production of climate security (see e.g. Solana 2008: 10), as it increases resilience against major disruptions through natural disasters in its southern vicinity. MENA partner countries, however, do not need to buy into this rationale, as the improvement in civil protection capacities for them is valuable on its own. Within the framework of the EMP, Euro-Mediterranean cooperation on civil protection was incorporated as a confidence-building measure (Bremberg 2010a: 144). Funded by the EU, two subsequent programs, including training, information sharing, the exchange of practitioners and experts as well as technical assistance, were implemented within the EMP.[8] The rationale behind these initiatives was basically to bypass the blockades of regional cooperation in the security field due to existing conflicts and antagonisms. Technocratic cooperation of experts should help to overcome divisions by establishing common routines, perceptions and norms. And indeed the programs have been successful in establishing a network of experts and practitioners in the field of disaster management (Bremberg 2010b: 182–183).

The UfM in 2008, then, made civil protection one of its priority areas. As a consequence in 2009 the 'Prevention, Preparedness and Response to Natural and Man-Made Disasters' (PPRD South) program was launched. One of the aims of the PPRD South was defined as "bringing our Candidate Countries of the Mediterranean region closer to the European Union Civil Protection Mechanism" (Füle 2012).[9] The PPRD sought to externalize the established cooperation mechanisms to Europe's southern vicinity without having to provide all the resources and personnel itself. As Bremberg (2010a: 146) puts it: "Euromed civil protection is increasingly being shaped as the external, albeit highly interconnected, dimension of EU civil protection."

Thus, in the late 2000s, cooperation on civil protection emancipated itself from being simply a confidence building measure to being a critical political asset in itself. The climate change and security discourse had a particular impact on this emancipation of civil protection (see Brauch 2012: 779). For example, the European foreign ministers in 2007 identified a: "rising trend in vulnerability to natural and man-made disasters in many parts of the Mediterranean" (European Commission 2007a), which would require increased and coordinated efforts in civil protection. The joint declaration of the UfM summit on civil protection in Paris in 2008 acknowledged that:

> The global landscape is littered with examples of the devastation caused by man-made and natural disasters. The effects of climate change are evident for all. The Mediterranean region is particularly vulnerable and exposed to such disasters.
>
> (UfM 2008: 19)

In line with the differential framing of the Mediterranean outlined above, climate change and security discourse problematized the vulnerability and limited political coping capacity in the MENA countries as a risk for the EU. Against the backdrop of this rationale, the externalization of its civil protection mechanism to promote resilience in its southern neighborhood is part of a preventive security policy (Bulley 2013: 265). Externalization of civil protection mechanisms, from this perspective, means governing climate disaster-related risks in the Mediterranean *at a distance* by increasing the preparedness and resilience of the affected (Joseph

2013: 43). And in line with such a resilience discourse, civil protection is increasingly rearticulated as climate adaptation:

> Neighbouring countries should be encouraged and supported in analysing impacts, risks, vulnerabilities and suitable responses and in including adaptation in their development plans. Actions with neighbours should build on strengthening of existing cooperation, dialogues and processes, in particular in the context of the European Neighbourhood Policy (ENP).
>
> (European Commission 2007b: 23)

Following a resilience rationale, the EU does not seek to run civil protection operations itself but rather to empower local forces and practitioners in the MENA region.

Securitization logic II: climate change and security as translation

It is not the case that MENA actors would simply be passive receivers of the efforts of different European actors to spread discourses of climate change and security or practices to deal with this threat in the southern Mediterranean. This becomes obvious when we turn to a second securitization logic, which I have called *securitization of climate change as translation.* This term refers to processes in which discourses on other political issues in the Mediterranean are *rearticulated* in relation to climate change and security. In terms of the securitization approach used in this book, this implies that already existing (transnational) discourse coalitions take up climate change and security into their repertoire of storylines. This holds particularly true for three discourses: illegal migration, desertification and water conflicts.

Climate change, home affairs and transit states

The first translation mechanism I would like to present is the reframing of a homeland security discourse in relation to the problem of CIM. A homeland security – or in EU terminology, *justice and home affairs* (Bicchi 2006: 201) – discourse deals with the multiple dangers to EU citizens in an interlinked and seemingly ever more complex world: transnational terrorism, organized crime, the spread of disease, human and drug trafficking, and illegal immigration. If one takes a closer look at European – or better transatlantic – articulations of climate security implications in the southern Mediterranean, it becomes clear that most do not deal with interstate war or conflict but with human security problems that could grow into homeland security concerns for the EU (e.g. Brown and Crawford 2009: 25–26; Busby *et al.* 2010: 5–6; NIC 2009). At the heart of this *climate change and home affairs discourse* lies the figure of the climate change migrant (Hartmann 2014: 139).

A lot has been written on the securitization of migration within the European Union and related critiques of a Fortress Europe and excessive practices of control and repression especially at the EU's southern borders. The establishment of FRONTEX and its operations in the Mediterranean as part of the securitization of migration has been likewise studied and criticized (Léonard 2010: 232–233; Neal 2009: 334). In recent years North African countries have played an increasing role in the detention of illegal migrants from sub-Saharan African countries. Countries that, from a European position, function as transit states,[10] such as Libya and Morocco, have willingly played the illegal migration card to improve relations with the global North and distance themselves from the poor South (White 2011: 120). Through bilateral cooperation agreements, e.g. between Spain and Morocco in 2004, or between Italy and Libya 2003 (Lutterbeck 2006: 72), southern Mediterranean countries became increasingly incorporated into the EU's approach of policing illegal immigration.[11] In the established institutions, practices and objects – such as walls, fences, detention camps, patrolling boats – of migration management in the region, the spatial concept of the southern Mediterranean as a buffer zone finds it material expression.

The figure of the illegal migrant represents a nodal point in the whole discourse on justice and home affairs. Metaphorically expressed in terms of the container model – as a 'carrier' of dangerous diseases, radical ideas, drugs or social contacts to suspicious and uncontrollable groups outside the EU – illegal immigration represents the channel through which dangers could enter the political 'container' of the Schengen space (see e.g. Maas and Tänzler 2009: 29). The figure of the climate refugee takes up a similar, yet slightly different, role. On the one hand, climate refugees coming from or through the region, unlike the illegal migrants, are not constructed as a national security but rather as a humanitarian or human security concern (see Methmann and Rothe 2014). Yet as climate refugees stem mainly from the most vulnerable and thus poorest regions, they are attributed with the same stigmatizing characteristics: poorly educated, impoverished, potentially criminal, etc. as expressed in this example: "Cross-border smuggling, human trafficking, and other illegal activity are likely to increase with greater migration and climatic stress" (NIC 2009: 32). What is more, climate change is anticipated to massively increase the number of migrants heading towards Europe. This is expressed in the following fictive future scenario from the WBGU report:

> Towards the middle of the 21st century the northwards migration of predominantly young men from the countries of the Sahel takes on the proportions of a "Völkerwanderung," i.e. mass migration such as was seen in Europe during the Dark Ages.
>
> (WBGU 2009: 125)

Lastly, as most climate refugees are going to stay within Africa, they might become a further source of instability in already fragile areas as they are "increasing the pressure on resources such as shelter, water, food, and energy" (Tänzler *et al.* 2010: 32). Hence, even if the climate migrant does not become an illegal migrant heading towards EU territory, he might become dangerous by triggering conflict in the southern neighborhood of the EU. For example the already quoted WBGU scenario continues with predicting that:

> The year 2020 sees the start of serious social and political destabilization in all the countries affected by this migration. The rural exodus caused by desertification and drought leads to further urbanization; as a result the urban slum areas expand further and their population density is very high. The situation of economic hopelessness generates enormous potential for political destabilization [...]. This creates a breeding ground for the further radicalization and spread of extremist religious movements.
>
> (WBGU 2009: 125)

This rationale discursively links CIM migration in the region to Islamist terrorism: Many areas in which stronger violent Islamist groups exist are particularly threatened by the negative impacts of climate change and less able to cope with them institutionally. In the absence of sufficient public institutions "popular discontent could be channeled by extremist groups" leading to "increased sympathy for Islamist organizations" (Maas and Tänzler 2009: 5–6).

The examples presented in this chapter clearly show how the problem of illegal migration and related 'homeland' security threats are increasingly rearticulated in the light of climate security (Phone interview with a Geneva-based environmental security expert, September 15, 2011). This *logic of translation* is not restricted to a European context. Rather, we can notice that North African transit states like Morocco are also self-actively playing the climate-induced migration card: "It [Morocco, author's note] did so to curry favour with European officials, establish its claims on the Western Sahara, and strengthen the status of its own émigrés abroad" (White 2011: 118). For example, the Moroccan King Mohammed VI and members of his government in past years repeatedly articulated the linkages between environmental problems in the region and security threats such as climate-induced migration. In doing this, they explicitly adopted neo-Malthusian storylines that link population pressures, environmental degradation and potential security threats such as conflicts and migration (see Chapter 5).[12] The repeated problematization of environment and security linkages, e.g. at the 2009 EU–Africa Summit in Lisbon, can be seen as part of "Morocco's self-positioning as a key player in migration management" (White 2011: 120). Hence, while in the Mediterranean there is no consent on climate change being a security issue, illegal migration is commonly articulated as a shared concern for both sides of the Mediterranean (MedSec 2009: 25). And as White's study, for example, shows, the actors in this coalition are increasingly rearticulating the threat of migration in terms of climate change and security.

Desertification, land degradation and climate change

The quotes of Moroccan officials in the previous section already point to a second environmental security discourse that has produced some resonance in the southern Mediterranean in the past: the threat of desertification. As I have outlined in more detail in Chapter 4, as early as the 1990s the problem of desertification has been linked to potential security threats such as migration or violent conflict. What is more, the discourse produced considerable resonance in the southern Mediterranean, probably as the storylines corresponded with its own experience of the West Saharan conflict. This problematization of deserts as dangerous spaces is taken up in discourses on climate change and security. Climate change and security discourse now translates the problem of desertification into an impact of climate change (e.g. WBGU 2009: 134). Translated into a climate impact, the desert as a risky space retained a central position in the discourse on climate security. There are at least two important reasons for this prominence.

The first reason refers to the prominent imaginary of the desert as an empty, uncontrolled and ungoverned space (e.g. Busby *et al.* 2010: 13). In this imaginary the desert becomes a source of danger, as it serves as a safe haven for terrorist and warlord groups (Hartmann 2010: 240–241). The empirical material for example refers to the Algerian terrorist group *Groupe Salafiste pour la Prédication et le Combat*, which in 2007 renamed itself as *Organisation al-Qaïda au Maghreb Islamique (Al-Queda Organization in the Islamic Maghreb: AQIM)*. The group – just like factions in Tunisia and Morocco – has since 2002 conducted a series of terrorist attacks in Algeria but recently also expanded its operational area to Saharan countries such as Niger and Mali. Several kidnappings of foreign tourists and businessmen can be attributed to the activities of the AQIM. Climate change, it is feared in this discourse, will on the one hand drive ever more Africans into the arms of radical groups due to migration, scarcity and related social grievances (e.g. Brown and Crawford 2009: 25). On the other hand, processes of desertification will provide such groups with ever more space to hide and organize themselves (NIC 2009: 32).

The second, related reason is that the desertification storylines perfectly resonate with Western stereotypes stemming from a *Mediterranean degradation narrative* which links land change to water stress and overpopulation (Phone interview with a member of the European External Action Service, November 25, 2011). As Davis (2010), for example, argued, the environmental imaginaries of the MENA region have played an important role in the European colonial project in that region. Environmental discourse on the southern Mediterranean from the beginning framed it as a marginal ecological space, deteriorated by the irrational use of resources by local populations – Davis has called this *environmental orientalism* (Davis 2010: 657). The indigenous mismanagement – following these Western narratives – produced a series of grave ecological problems such as desertification through deforestation and overgrazing of land or water scarcity due to over-irrigation. In the past, these storylines have been used to legitimize different colonial practices such as land grabbing, the introduction of industrialized agriculture or techno-political projects such as the construction of the Aswan Dam in Egypt in 1902 (Mitchell 2002). In the present climate change and security discourse, the narrative prevails in articulations that link land degradation, population and potential conflict: "if deserts spread there is less arable land and people are going to migrate" (Phone interview with an Amsterdam-based climate and energy activist, September 26, 2011).[13] Here an antagonist security discourse that problematizes economic and political inferiority in the South finds its way into climate security discourse.

Water conflicts and climate change

The essential characteristic of the climate change and security discourse is not only that it makes other discussed threats worse but also that it makes them *complex and hence unpredictable* (Grove 2010: 547). Let me exemplify this point by introducing a further political problem that translated into climate security language: the threat of *water conflicts* (e.g. Shiva 2002). The narrative of water conflicts had been popularized even before a major securitization of climate change took place (see e.g. Gleick 1993). Water war storylines have for example been made popular by movies such as *Blue Gold* by Sam Bozzo, and semi-scientific non-fiction books such as *World War Water* (Mostofizadeh 2012). As becomes apparent from the latter's description, water in this narrative is first constructed in economic terms as a scarce and valuable natural resource: "The next world war will not be over oil or energy, but it will most likely be over water" (Mostofizadeh 2012). Water is brought into an equivalential relationship to other natural resources that are usually regarded as being embattled. Just as with storylines on a war for oil, water is sometimes presented as the *hidden cause* behind political conflicts. This holds especially true for the Middle East conflict, which in some literature is related to water (e.g. Zeitoun 2008).

Having outlined the water security narrative in very general terms, let us turn to the specific water conflict narratives in climate change and security discourse on the Mediterranean. First of all, it is important to note that almost all discursive articulations stress the fact that the issue is contested in scientific literature and that water scarcity under specific conditions could even foster cooperation rather than causing conflict (Brown and Crawford 2009: 20). On the one hand, due to the high number of transnational waters such as the Nile, the Tigris, the Euphrates and the Jordan, the MENA region in general is perceived as a highly vulnerable region with regard to water security (Phone interview with a Beirut-based climate activist, September 16, 2012). On the other hand, articulations in the climate change and security discourse on the Mediterranean do not conceal that the region at the same time provides the most impressive examples of a peaceful resolution of water disputes.[14]

However, the situation changes when dangerous climate change enters the stage. This is because it acts as a *game changer*, which changes the nature of water-related risks in the region. This is exemplified in the following quote from the EU-funded (2010–2013) 'Climate Change, Hydro Conflicts and Human Security' project CLICO,[15] which explicitly argues drawing simple correlations between water stress and conflict:

> The reality is much more complex. The nature of the *hazards*, for instance, influences the type of impacts and insecurity produced [...].Water scarcity but also extreme surpluses such as floods or flash floods and the timing of hazards matter [...]. Combinations of hazards such as droughts followed by extreme precipitation resulting in flooding can have particularly far reaching and long-term impacts.
>
> (Kloos *et al.* 2013: 32)

As the discourse on climate change-related security risks unfolds, the focus is shifted away from full-scale interstate disputes over shared waters to the impact of so-called hydrological *extreme events*: "climate shocks [...], rather than absolute scarcity of resources like water and land, are thought to be the most likely way climate change could contribute to conflict" (Busby *et al.* 2010: 3). This also represents a discursive shift in the problematization of water security in the context of climate change. Now, it is not water scarcity which poses a threat to security any more but the focus is on 'Hydrohazards' which are defined as "threats (perturbations and stresses) arising from the water system such as floods, droughts, land slides" (Kloos *et al.* 2013: 40). While water conflict storylines originally dealt with conflicts over existing, slowly decreasing water resources, the threats presented in the quote lie in an uncertain future. What is crucial is hence the *temporality* of the water security threats. This shifts from a focus on gradual long-term changes to the non-linear and unpredictable combination of chronic scarcities with surprising shocks.

Alongside the temporal dimension, there is a rearticulation of the spatial dimension of the water security threat. Instead of presenting the whole Mediterranean region as being threatened by water scarcity and its conflict potentials, the discourse now increasingly paints a more fragmented picture of diverging local and regional vulnerabilities (Kloos *et al.*

2013: 8). This is reinforced by the fact that computer models of regional climate change impacts in terms of regional precipitation patterns, the frequency of droughts, sea-level rise and the likelihood of disasters have hitherto been rather unreliable and uncertain (Interview with a climate security researcher, Hamburg, November 17, 2011). While there is high certainty that the region is hit hard by dangerous climate change, there is likewise a high uncertainty about the concrete forms that this harm is going to take and about the affected areas (Kloos *et al.* 2013: 23).

The situation becomes even worse when we broaden our perspective to include the problems that are linked with this issue. A nice visualization of this complexity is presented in Figure 6.5, which depicts the interlinkages between water issues and other problems. What becomes apparent from this figure, is that water security, or the availability of freshwater resources, is articulated alongside other potential harms or threats, each of which is subsumed under the broader phenomenon of climate change. A growing scarcity of drinking water is projected to absorb huge amounts of energy (for example for desalination), putting constraints on energy resources, which are already becoming scarcer; the scarcity of groundwater also challenges agricultural irrigation, which has to cope with rising temperatures and droughts undermining the food security of MENA populations (Figure 6.5). Water scarcity interacts with these other security problems in a complex and non-linear way: "Due to the complexities within the natural system and its interlinkages to the social, economic and political spheres, a highly complex nexus has evolved that connects *climate change, water conflicts and human security*" (Kloos *et al.* 2013: 7).

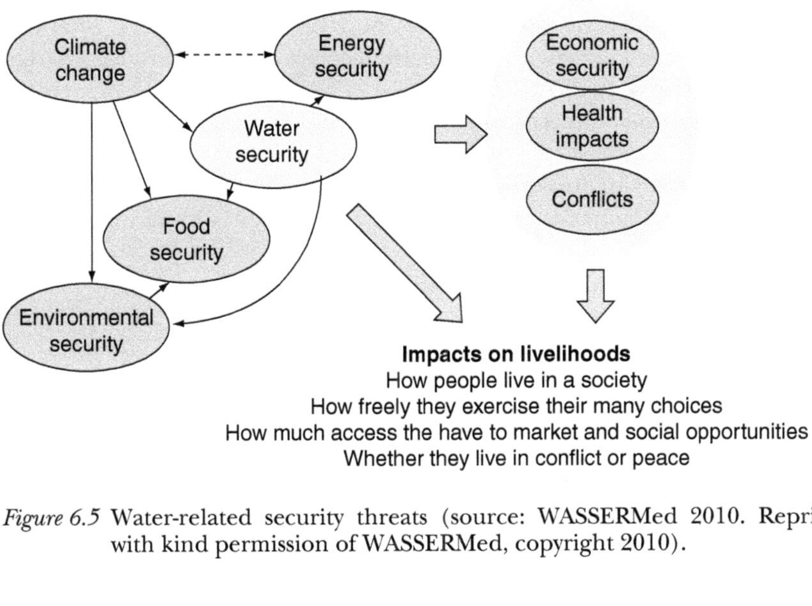

Figure 6.5 Water-related security threats (source: WASSERMed 2010. Reprinted with kind permission of WASSERMed, copyright 2010).

We can thus say that water conflict storylines have been rearticulated in relation to an epistemology of complexity that characterizes climate security discourse (Mayer 2012: 172–174). The discourse shifted away from a geopolitical take on water conflicts focusing on conflicts to slowly diminishing water resources between political enemies and a problematization of vaguely defined dangers in terms of vulnerability, risk and resilience. In this complexity discourse, the relevant reference objects are not political actors, i.e. countries, or communities, anymore. Instead the objects of climate insecurity are now understood as *socio-ecological systems* (SES): "An SES is based on a bio-geo-physical unit and its associated social actors and institutions. SESs are complex and adaptive and delimited by spatial or functional boundaries surrounding particular ecosystems and their problem context" (Kloos *et al.* 2013: 32).

And the EU's policy approach that follows from such a problematization of socio-ecological vulnerability is one of promoting the resilience of such socio-ecological systems to climate change impacts. The unpredictability of climate change impacts in the MENA region combined with the complexity of the political situation in these countries poses serious obstacles both to a centralized government and to a market-driven approach to managing climate risks. The logical response is to concentrate on the resilience of populations at the local level. A perfect example is the program 'Social and ecological resilience in the face of climate change in the Mediterranean Region (SEARCH)' run by the EU from 2010 until 2013. In the words of the program description, the EU's approach is not to develop a comprehensive framework to govern climate risks in the Mediterranean. Instead "SEARCH concentrates on giving the local community and other stakeholders the skills and knowledge they need to assess the impact of climate change on a day-to-day basis" (ENPI-Info 2015). Instead of meeting their historic responsibility and providing massive financial assistance, European countries restrict themselves to providing knowledge and ideational support. SEARCH "increases the knowledge and capacity of the local community with regard to climate change adaptation planning and tools" and establishes "regional knowledge networks" (ENPI-Info 2015). And instead of top-down government interventions it "uses joint participatory approaches to plan for climate change adaptation locally" (ENPI-Info 2015).

Complex entanglements

When we compare the different discourses that are translated within a discourse on regional climate security implications, there is one connecting theme: a growing complexity or unpredictability, combined with limited governmental capacity in the southern Mediterranean to cope with this contingency. Uncertain *hydro-futures* link up with unpredictable flows of *CIM*, while spreading *deserts* and uninhabitable land functions as a safe haven for extremists. For the EU such "systemic risks" (EEAS 2011) are particularly problematic due to the spatial proximity of the southern Mediterranean to its borders. What makes climatic changes dangerous is the speed and the non-linearity of its impacts.[16] For example, climate change, it is feared, will not only increase but also change the nature of migration. A refugee who moves from point A to B is not a problem for the state – he or she can easily be treated within the existing institutions to manage migration. Migration from North Africa to Europe in the past followed relatively stable patterns (Fargues 2009: 19). Climate-induced migration is different in this respect: where disasters might occur next is not predictable, just as the preferred destination of people on the move is unknown. As several articulations in the discourse acknowledge, people displaced from climate-related disasters will probably not head towards industrialized countries (e.g. Busby *et al.* 2010: 7). People in general want to stay close to their homes, and could use temporary migration as a means to adapt, for example to climate-related disasters (Busby *et al.* 2010: 6). As such climate migrants are not graspable, sizable or representable for the institutions of the state.

CIM becomes particularly dangerous when it comes together with a second problem – namely 'state failure.' The absence of state institutions is seen as problematic because it leads to the emergence of smooth spaces: spaces in which dangers can freely circulate and thereby accelerate and reinforce each other. The securitization of desertification presented above has shown that climate change is likely to produce more smooth spaces, i.e. spaces in which the organizations of the state rarely exist. If climate change-induced migration hits such spaces, there is the danger that flows become accelerated or that they feed into political movements (NIC 2009: 32). According to Betsy Hartmann (2014: 774), climate security thereby feeds into recent rethinking in defense policy towards counterinsurgency and military action against nomadic tribes, for example in the US Defense Department's Trans-Saharan Counter Terror Initiative. However, it is not only the deserts but also an imagined, "chaotic urbanization" (MedSec 2009: 24) as a result of climate change that can produce smooth spaces: "Ghetto-like environments" (NIC 2009: 15), in which dangerous diseases, subjects or ideas can freely circulate and accelerate, as "the urban slums threaten to become lawless areas" (WBGU 2009: 125).

As a result the climate change and security discourse legitimizes an increased surveillance and monitoring of (North) Africa through the intelligence community (Hartmann 2014: 775). A perfect example to illustrate this trend is the European Union's Earth Observation Program, called Copernicus. Copernicus emerged out of the older Global Monitoring of Environment and Security program in 2014 and represents a massive EU initiative to centralize the earth observation capacities that exist in single member states and to build up its own European capacities by installing up to six earth observation satellites (so called SENTINALS) until 2020.

With an overall financial budget of 4.29 billion euros, Copernicus is the most ambitious ecological remote sensing program since the launch of NASA's Landsat in 1972. Copernicus is often presented as a mere environmental project including monitoring services on different earth systems such as the atmosphere, marine environment, or land surface. However, a crucial reason for the Copernicus's political appeal actually stems from its non-environmental services, which include disaster prevention, the monitoring of conflict areas and the surveillance of the Mediterranean as a crucial part in the establishment of a comprehensive maritime and border surveillance system (Levy 2011). And in these security related services of Copernicus, G-NEXT and G-Sextant, which are currently still in a pre-operational test phase, the influence of climate security rationales is conspicuous. For example, G-Sextant includes a service on informal settings in crisis regions, which includes monitoring refugee camps and informal settlements. The link to the climate security discourse and its problematization of ungoverned spaces becomes obvious in the service description:

> At present, an accelerated process of urbanisation is taking place in many different areas, particularly across Asia and Africa [...].This may lead to a widespread sense of insecurity amongst the population, increased crime rates, and ultimately riots and political instability. EO [Earth Observation, author's note] products allow the monitoring of urban growth and the identification of potential hot spots that decrease urban resilience.
>
> (G-SEXTANT 2014)

In a different program, G-Sextant seeks to provide early warning of looming conflicts in environmental hotspot regions. The satellite images provided by Copernicus's pilot services on security deliver geoinformation about phenomena, which are closely related to such a neo-Malthusian script: natural resources, land-use (change), population size and population growth. Not surprisingly the pilot services on environmental conflicts and migration concentrate exclusively on African countries, such as for example Zimbabwe or the DR Congo (G-SEXTANT 2014: 7). The Copernicus satellites Sentinel 1 and Sentinel 2 are perfectly adapted to cover the required data for such services. Apparently technical questions, such as the inclusion of different spectral bands, thus become inherently political, as what the installed satellites can see goes back to contested political storylines such as that of climate change and conflict.

Certainly, the securitization of climate change is not the single cause of the establishment of Copernicus. But it is justified to say that the scenario of Europe's southern backyard descending into chaos has created an environment in which joint monitoring of the environment, disasters and migrants appears reasonable.

Climate security enters the governmental landscape of EU region building

In this chapter I was able to show that the region of the southern Mediterranean played a crucial role for the European discourse on climate change and security due to the region's proximity and its function as a southern flank, gateway or buffer zone to the African continent. Throughout the chapter I revealed different channels through which the climate security discourse trickled down into the regional politics in the Mediterranean. These can be found in the cooperation on civil protection and disaster resilience, in the field of policing migration, as well as in an increasing monitoring and surveillance of Europe's southern environment. None of these political measures has been instituted by a securitization of climate change alone. And yet, it is fair to say that none of them would have had such political significance if the (southern) Mediterranean had not been constructed as a hotspot for the security implications of climate change.

Unlike as suggested by the Copenhagen School approach, the securitization process here does not bring about any political change in the sense of the adoption of extreme measures. On the contrary, the emerging climate security discourse on the Mediterranean clearly had a status quo bias in that it stabilized the existing governmental landscape in the region and provided further rationales for its preservation in the sense of a defensive hegemony, as outlined in Chapter 3. Instead of a coherent Mediterranean-wide climate security strategy, which outlines a clear political subject (the region) as well as a common governance object (the production of climate security) by clear-cut policy measures, there is a complex topology of different measures at different scales and different times which are loosely related to the climate security discourse. Such a fragmented approach is in line with the overall EU policy in the region after the Arabellions, which reacts to the complex and difficult realities in the region with a focus on single pragmatic and technocratic projects rather than having an overall strategy of regional integration (Horst *et al.* 2013). The resulting governmental landscape of climate security in the Mediterranean is characterized by political rationalities that oscillate from a cooperative security discourse and the notion of the Mediterranean as a genuine region, through the notion of the southern Mediterranean as a buffer zone towards the problematization of MENA as Europe's chaotic neighborhood. The correlating technologies of power included both forms of neoliberal government at-a-distance as well as more repressive forms of biopolitical control. Climate change and security discourse on the Mediterranean sits comfortably within this governmental landscape and hence legitimizes its further continuance. First, the notion of the Mediterranean as a region facing the common threat of climate change provides the legitimacy for continued and increased European intervention in and cooperation with its southern neighborhood. Second, practices of producing resilience in MENA countries should increase the adaptive capacities in such countries and thus prevent them becoming a risk for Europe. Finally, increased control and monitoring of the Mediterranean Sea, ports, cities, desert areas and the movement of people function as a safeguard in case those soft measures of governing at a distance prove insufficient.

Notes

1 It is important to note that neither the Mediterranean nor MENA represents a region that is a natural given (Jünemann *et al.* 2013). In using the term MENA or Mediterranean region, I refer to the political space that is constituted by the respective signifiers.

2 To be precise here: the term *climate security discourse on the Mediterranean* means discursive articulations that *refer to* the region and not to voices of actors from within the Mediterranean, as the alternative formulation *climate security discourse in the Mediterranean* would imply.

3 A few passages of this chapter have already been published (in Methmann and Rothe 2014). Reprinted with permission of Taylor and Francis.

4 The establishment of the External Borders of the Member States of the European Union (FRONTEX) in 2004 is a good example of such surveillance. For example, its Rapid Border Intervention Teams run operations in the Mediterranean Sea to detect and detain illegal immigration to the EU by water (Léonard 2010: 241).

5 The IPCC did not account for the Mediterranean as a region either. Rather, the IPCC took Africa, including the North African countries, as one region, just as it did with Europe (Watson *et al.* 1997: 6–11). The Middle East was assessed together with the Central Asian arid region (Watson *et al.* 1997: 7).

6 Cartography and the practice of mapping is a practice or technology of power that produces a particular type of knowledge (Harley 2009).

7 There are further examples similar to the MedSec case. In June 2012, for example, the OSCE hosted a high level 'Participatory Workshop on Environment and Security Issues in the Southern Mediterranean Region' (IES 2012), which also featured a mapping exercise similar to that of the Medsec initiative.

8 These were the 'Pilot Programme for the Creation of a Euro-Mediterranean System of Mitigation, Prevention, and Management of Natural and Man-Made Disasters' running from 1998 to 2004 and the 'Euromed Bridge Programme for the Prevention, Reduction and Management of Natural and Man-Made Disasters' running from 2005 until the birth of the UfM (Bremberg 2010a: 144–145).

9 The *Community Mechanism*, established in 2001, is an outcome and sign of a rapid European integration process in civil protection from the early 2000s and culminated in its own article on disaster management in the Lisbon Treaty. The Community Mechanism provides the legal basis of joint European efforts in disaster prevention and relief in EU Member but also non-Member countries (Bremberg 2010a: 145–146).

10 For example expressed in this quote: "North African countries play a special role in migration movements on the continent: they are themselves migration destinations (for both internal and cross-border migrants), but at the same time they are a transit area for people from sub-Saharan Africa and Asia attempting to reach Europe (transit migration)" (WBGU 2009: 124).

11 This cooperation took different forms, from running joint patrolling operations in the Mediterranean Sea to readmission agreements and the establishment of detention camps in MENA countries (Lutterbeck 2006; White 2011–8).
12 For example, in 2009 Morocco hosted the meeting of the *International Union for Scientific Study of Population*, at which High Commissioner for Planning Lahlimi Alami stated: "Food security, desertification, farmland depletion and sea level rise are pressing issues for demographers to ponder [...] Located at a point where North meets South, Morocco feels very much concerned about migration issues" (Alami quoted in White 2011: 92).
13 Several authors have, for example, attributed the cascade of the 'Arabellions' to the negative impacts of climate change on the region (e.g. Johnstone and Mazo 2011). A climate change-driven water and food crisis, following this storyline, mixed with a socio-economic situation in which population growth and economic mismanagement had produced a large number of young unemployed and dissatisfied people: "a demographic 'youth bulge'" (NIC 2009: 13). In this fragile situation, then, a single event such as the self-immolation of a Tunisian vegetable hawker could become a trigger for a political wildfire in the whole region.
14 Despite the existence of fierce political antagonisms, for example, it was possible for Israel to sign water agreements with Jordan and to prevent violent conflicts over joint resources (Brown and Crawford 2009: 21).
15 The CLICO project was part of the CLIWASEC Cluster, 'Climate Change Impacts on Water and Security (in Southern Europe and neighbouring regions)' funded under the EU's 7th Framework Program for Research and Technical Development from 2010 until 2013 (CLIWASEC 2010).
16 Elsewhere, Chris Methmann and I have explained this through recourse to Deleuze and Guattari's interpretation of the nomadic war machine (Methmann and Rothe 2014: 173–174).

References

Adler, Emanuel and Beverly Crawford. 2004. "Normative power: the European practice of region building and the case of the Euro-Mediterranean partnership (EMP)." Berkeley CA: Institute of European Studies.

Bauer, Patricia. 2015. "The European Mediterranean policy after the Arab Spring: beyond values and interests." *Middle East Critique* 24 (1): 27–40.

Bicchi, Federica. 2006. "'Our size fits all': normative power Europe and the Mediterranean." *Journal of European Public Policy* 13 (2): 286–303.

Brauch, Hans G. 2010. "Climate change and Mediterranean security: international, national, environmental and human security impacts for the Euro-Mediterranean region in the 21st century." In *IEMed Papers No 9*, edited by IEMed. Barcelona: IEMed.

Brauch, Hans G. 2012. "Policy responses to climate change in the Mediterranean and MENA region during the Anthropocene." In *Climate change, human security and violent conflict*, edited by Jürgen Scheffran, Michael Brzoska, Hans Günter Brauch, Peter M. Link and Janpeter Schilling, 719–794. Berlin/Heidelberg: Springer.

Bremberg, Niklas. 2010a. "Civil protection: the new dimension of Mediterranean security." In *Mediterranean Yearbook 2010*, edited by European Institute of the Mediterranean, 144–147. Barcelona: IEMed.

Bremberg, Niklas. 2010b. "Security, governance and community beyond the European Union: exploring issue-level dynamics in Euro-Mediterranean civil protection." *Mediterranean Politics* 15 (2): 169–188.

Brown, Oli and Alec Crawford. 2009. "Rising temperatures, rising tensions." In *Climate change and the risk of violent conflict in the Middle East.* Winnipeg: International Institute for Sustainable Development.

Bulley, Dan. 2013. "Producing and governing community (through) resilience." *Politics* 33 (4): 265–275.

Busby, Joshua, Kaiba White and Todd Smith. 2010. *Mapping climate change and security in North Africa.* Washington, DC: the German Marshall Fund of the United States.

Center for Naval Analysis. 2007. "National security and the threat of climate change." Last Modified 2007. Accessed May 4, 2015. https://www.cna.org/sites/default/files/National Security and the Threat of Climate Change – Print.pdf.

CLIWASEC, Climate induced changes on water and security. 2010. "Projects." Accessed May 4, 2015. www.cliwasec.eu/projects/projects.php.

Council, the Council of the European Union. 2009. "Joint progress report and follow-up recommendations on climate change and international security of work." 16645/09, Brussels.

Crampton, Jeremy W. 2001. "Maps as social constructions: power, communication and visualization." *Progress in Human Geography* 25 (2): 235–252.

Davis, Diana K. 2010. "Power, knowledge, and environmental history in the Middle East and North Africa." *International Journal of Middle East Studies* 42 (04): 657–659.

Diez, Thomas. 2005. "Constructing the self and changing others: reconsidering normative power Europe." *Millennium-Journal of International Studies* 33 (3): 613–636.

Dimitrovova, Bohdana. 2008. "Re-making of Europe's borders through the European neighbourhood policy." *Journal of Borderlands Studies* 23 (1): 53–68.

EEAS, European Union External Action Service. 2011. "Joint reflection paper: towards a renewed and strengthened EU climate diplomacy." Accessed May 4, 2015. http://eeas.europa.eu/environment/docs/2011_joint_paper_euclimate_diplomacy_en.pdf.

Elasha, Balgis Osman. 2010. "Mapping of climate change threats and human development impacts in the Arab region." In *UNDP Arab development report–research paper series,* New York: UNDP Regional Bureau for the Arab States.

ENPI-Info, European Neighbourhood Info Centre. 2015. "Social and ecological resilience in the face of climate change in the Mediterranean Region (SEARCH)." Accessed May 5, 2015. www.enpi-info.eu/mainmed.php?id=439&id_type=10.

EU Security Strategy. 2003. *A secure Europe in a better world: the European security strategy,* drafted by EU High Representative Javier Solana. Brussels: European Council.

European Commission. 1994. "Communication from the Commission to the Council and the European Parliament: strengthening the Mediterranean policy of the European Union of Work." Brussels: COM(94) 427 final.

European Commission. 2007a. "Communication from the Commission to the Council and the European Parliament on the preparation of the Lisbon Euro-Mediterranean Foreign Affairs Conference of Work." Brussels: COM(2007) 598 final.

European Commission. 2007b. "Green paper from the Commission to the Council, the European Parliament, the European Economic and Social Committee and the Committee of the Regions – adapting to climate change in Europe – options for EU action of Work." Brussels: COM(2007) 354 final.

European Commission. 2011a. "Joint communication by the High Representative of the Union For Foreign Affairs and Security Policy and the European Commission: a new response to a changing neighbourhood. A review of European neighbourhood policy of work." Brussels: COM(2011) 303.

European Commission. 2011b. "Joint communication to the European Council, the European Parliament, the Council, the European Economic and Social Committee and the Committee of the Regions: a partnership for democracy and shared prosperity with the southern Mediterranean of work." Brussels: COM(2011) 200 final.

Fargues, Philippe. 2009. *Mediterranean migration report 2008–2009.* San Domenico di Fiesole: CARIM, Robert Schuman Centre for Advances Studies, European University Institute.

Füle, Štefan. 2012. "The future prospects for Euro-Mediterranean co-operation in the field of civil protection: speech of the European Commissioner for Enlargement and Neighbourhood Policy, Brussels: European Commission.

G-SEXTANT, Geospatial Intelligence Services in Support of EU External Action. 2014. "Humanitarian crises." Accessed May 5, 2015. http://externalaction.security-copernicus.eu/projects-overview/g-sextant/humanitarian-crisis.

Gleick, Peter H. 1993. "Water and conflict: fresh water resources and international security." *International security*: 79–112.

Grove, Kevin J. 2010. "Insuring 'our common future?' Dangerous climate change and the biopolitics of environmental security." *Geopolitics* 15 (3): 536–563.

Harley, J Brian. 2009. "Maps, knowledge, and power." In *Geographic Thought–A praxis perspective*, edited by George Henderson and Marvin Waterstone, 129–149, New York: Routledge.

Harley, John Brian. 1989. "Deconstructing the map." *Cartographica: The international journal for geographic information and geovisualization* 26 (2): 1–20.

Hartmann, Betsy. 2010. "Rethinking climate refugees and climate conflict: rhetoric, reality and the politics of policy discourse." *Journal of International Development* 22 (2): 233–246.

Hartmann, Betsy. 2014. "Converging on disaster: climate security and the Malthusian anticipatory regime for Africa." *Geopolitics* 19 (4): 757–783.

Horst, Jakob. 2013. "Free trade, development and authoritarianism in a rentier economy: intentions and reality of the EU's logic of action in Algeria." In *Euro-Med relations after the Arab Spring: persistence in times of change*, edited by Jakob Horst, Annette Jünemann and Delf Rothe, 79–102. Farnham: Ashgate.

Horst, Jakob, Annette Jünemann and Delf Rothe. 2013. "Conclusion." In *Euro-Med relations after the Arab Spring: persistence in times of change*, edited by Jakob Horst, Annette Jünemann and Delf Rothe, 225–232. Farnham: Ashgate.

Howden, Daniel. 2007. "Diplomats warned that climate change is security issue, not a green dilemma." *The Independent*, December 6, 2007, 36.

Huysmans, Jef. 2000. "The European Union and the securitization of migration." *JCMS: Journal of Common Market Studies* 38 (5): 751–777.

IES, Institute for Environmental Security. 2012. "Participatory workshop on environment and security issues in the Southern Mediterranean region, Amman, Jordan, June 18–22, 2012.

Jeandesboz, Julien. 2007. "Labelling the/'neighbourhood/': towards a genesis of the European neighbourhood policy." *Journal of International Relations and Development* 10 (4): 387–416.

Johnstone, Sarah and Jeffrey Mazo. 2011. "Global warming and the Arab Spring." *Survival* 53 (2): 11–17.

Joseph, Jonathan. 2013. "Resilience as embedded neoliberalism: a governmentality approach." *Resilience* 1 (1): 38–52.

Jünemann, Annette. 2009. "Zwei Schritte vor, einer zurück. Die Entwicklung der europäischen Mittelmeerpolitik von den ersten Assoziierungsabkommen bis zur Gründung einer Union für das Mittelmeer." In *Der Mittelmeerraum als Region*, 26–59. Tübingen: Europ. Zentrum für Föderalismus-Forschung.

Jünemann, Annette, Florian Kühn, Jakob Horst, Eva-Maria Maggi and Delf Rothe. 2013. "Introduction: setting up the logics of action framework." In *Euro-Med relations after the Arab Spring: persistence in times of change*, edited by Jakob Horst, Annette Jünemann and Delf Rothe, 1–20. Farnham: Ashgate.

Kloos, Julia, Niklas Gebert, Therese Rosenfeld and Fabrice Renaud. 2013. *Climate change, water conflicts and human security: regional assessment and policy guidelines for the Mediterranean, Middle East and Sahel.* Bonn: United Nations University, Institute for Environment and Human Security (UNU-EHS).

Korkmaz, Visne. 2008. "Constructing the Mediterranean in the face of new threats: are the EU's words really new?" *European Security* 17 (1): 141–160.

Kurki, Milja. 2011. "Governmentality and EU democracy promotion: the European instrument for democracy and human rights and the construction of democratic civil societies1." *International Political Sociology* 5 (4): 349–366.

Latour, Bruno. 1993. *We have never been modern.* Cambridge, MA: Harvard University Press.

Léonard, Sarah. 2010. "EU border security and migration into the European Union: FRONTEX and securitisation through practices." *European Security* 19 (2): 231–254.

Levy, Adam. 2011. "Intervention – remote sensing as remote control? A political geography of EU border surveillance." Accessed May 4, 2015. http://antipode-foundation.org/2013/01/29/intervention-remote-sensing-as-remote-control-a-political-geography-of-eu-border-surveillance/.

Lutterbeck, Derek. 2006. "Policing migration in the Mediterranean: ESSAY." *Mediterranean politics* 11 (1): 59–82.

Maas, Achim and Dennis Tänzler. 2009. *Regional security implications of climate change. A synopsis.* Berlin: Adelphi research.

Malmvig, Helle. 2004. "Cooperation or democratisation? The EU's conflicting Mediterranean security discourses." Copenhagen: DIIS Working Papers.

Mayer, Maximilian. 2012. "Chaotic climate change and security." *International Political Sociology* 6 (2): 165–185.

MedSec, Mediterranean Security Partnership. 2009. *Environment and security issues in the Southern Mediterranean.* Geneva: Zoï Environment Network.

Methmann, Chris and Delf Rothe. 2014. "Tracing the spectre that haunts Europe: the visual construction of climate-induced migration in the MENA region." *Critical Studies on Security* 2 (2): 162–179.

Mitchell, Timothy. 2002. *Rule of experts: Egypt, techno-politics, modernity.* Berkeley CA: University of California Press.

Mostofizadeh, Kambiz. 2012. *World war water.* Los Angeles: Mikazuki Publishing House.

Neal, Andrew W. 2009. "Securitization and risk at the EU border: the origins of FRONTEX" *Journal of Common Market Studies* 47 (2): 333–356.

NIC, National Intelligence Council 2009. *North Africa: the impact of climate change to 2030: geopolitical implications.* Washington, DC: NIC.

Öktem, Kerem. 2010. "Mediterranean frontiers. Borders, conflict and memory in a transnational world." In *The library of international relations, 46,* edited by Dimitar Bechev and Kalypso Nicolaidis, 13–34. London: Tauris Academic Studies.

Pace, Michelle. 2004. "The Euro-Mediterranean partnership and the Common Mediterranean strategy? European Union policy from a discursive perspective." *Geopolitics* 9 (2): 292–309.

Pace, Michelle. 2009. "Paradoxes and contradictions in EU democracy promotion in the Mediterranean: the limits of EU normative power." *Democratization* 16 (1): 39–58.

Shiva, Vandana. 2002. *Water wars: pollution, profits and privatization.* London: Pluto Press.

Smith, Karen E. 2005. "The outsiders: the European neighbourhood policy." *International affairs* 81 (4): 757–773.

Solana, Javier. 2008. *Climate change and international security.* Paper from the High Representative and the European Commission to the European Council. Brussels: European Commission.

Tänzler, Dennis, Achim Maas and Alexander Carius. 2010. "Climate change adaptation and peace." *Wiley Interdisciplinary Reviews: Climate Change* 1 (5): 741–750.

Traub-Werner, Marion. 2007. "Free trade: a governmentality approach." *Environment and Planning A* 39 (6): 1441.

UfM, Union for the Mediterranean. 2008. "Joint declaration of the Paris summit for the Mediterranean in Paris, July 13, 2008 of Work." Paris: CL08–090EN.

WASSERMed. 2010. "Water availability and security in Southern Europe and the Mediterranean: the project". Accessed May 6, 2015. http://wassermed.cmcc.it/index.php?option=com_content&view=article&id=56&Itemid=27.

Watson, Robert T., Marufu C. Zinyowera, and Richard H. Moss. 1997. *The regional impacts of climate change: an assessment of vulnerability.* Geneva: Intergovernmental Panel on Climate change (IPCC).

WBGU, German Advisory Council on Global Change. 2009. *Climate change as a security risk.* London: Earthscan.

White, Gregory. 2011. *Climate change and migration: security and borders in a warming world.* Oxford: Oxford University Press.

Wood, Denis and John Fels. 1986. "Designs on signs/myth and meaning in maps." *Cartographica: The International Journal for Geographic Information and Geovisualization* 23 (3): 54–103.

Zeitoun, Mark. 2008. *Power and water in the Middle East: the hidden politics of the Palestinian–Israeli water conflict.* London: IB Tauris.

7 The United Kingdom

Complexity, climate security and resilience

The UK is generally seen as a forerunner in the discourse on climate change and security. With regard to the political implications of this securitization process, however, there seems to be less consent. Some see clear political implications:

> I think when it comes to the political impact, the institutionalization of the discourse, you can see the clearest effects in Great Britain ... institutionally, because they have completely re-organized their governance system according to climate security. For example, by establishing a climate and energy department ... for example, through the climate change act or within the military and security sector.'
>
> (Interview with a German governmental advisor, Berlin September 14, 2011, author's translation)

Other experts who were asked about the policy implications of a climate change and security discourse in the UK painted a completely different picture: "I think that the real politics has remained pretty much the same. That may change in the coming years, but at the moment personally I don't see much of change" (Phone interview with a member of the UK MOD, September 18, 2012).

In this chapter I set out to explain how these diverging perceptions come about. On the one hand, I would argue that the question of political implications depends on what you look at. While it is difficult to attribute recent UK climate change legislation to the influence of a securitization process, one can clearly observe how the convergence of security and environmental discourse has changed the understanding and conceptualization of security policy in the UK. From this perspective, the apparent failure of securitization in the terms of the Copenhagen School – due to an absence of exceptional measures in climate policy – "is better understood as a 'successful climatization' of the security field" (Oels 2012: 185). The UK is a promising case to study a *climatization of security policy* because the relevant security and defense actors are situated at the national political level.

A second argument for why the evaluation of the policy implications of climate security discourse in the UK is so disparate is that there is nothing like a genuine governmental program on climate change and security. Rather, we can see the development of a complex *climate change and security assemblage.* Different parts of a UK climate security discourse coalition approached the issue through their established practices, rationales and technologies from very different fields and institutional backgrounds. While the practices of governmental actors from the Ministry of Defense (MoD), Department of International Development (DfID), and the Foreign and Commonwealth Office (FCO) were reoriented towards climate change, they were not subjected to a broader governmental logic.

A changing security landscape in the UK

The securitization of climate change in the UK took place within a broader reconceptualization process of UK security and defense policy. In the following I present some crucial features of this reconceptualization by reviewing the most important security policy documents at the national level: the National Security Strategies (NSS) from 2008 (Cabinet Office 2008) and 2010 (Cabinet Office 2010b) as well as the Strategic Defence Review 2010 (Cabinet Office 2010a).

At first sight, one could understand the reconceptualization of security in the UK as a shift from a geopolitical security approach towards biopolitical security and risk management. A first indicator that speaks to such a shift is the *multiplication* of phenomena that are considered relevant for security policy:

> Our national interest can be threatened by natural disasters, man-made accidents and by malicious attacks both by states and by non-state actors, such as terrorists and organised criminals. These risks have different impacts if they occur.
>
> (Cabinet Office 2010b: 25)

Especially visible is such an apparent shift through the growing importance of risk assessment practices in UK security policy. As Dunn-Cavelty and Hagmann argue, in contemporary UK security policy, it is not the political sovereign but rather the expert knowledge of risk managers that identifies security threats to the UK (Hagmann and Cavelty 2012: 83–84). The UK draws on different methods to anticipate the future, to identify patterns, to monitor and assess dangers and to establish risk profiles and probabilities – techniques that, according to Dillon and Lobo-Guerrero, are characteristic for a biopolitical security paradigm (Dillon and Lobo-Guerrero 2008). Decisive for the expansion of risk assessment and management in security policy was a new role for civil defense/civil protection agencies after the end of the Cold War (Hagmann and Cavelty 2012: 83).

With the diminishing threat of a foreign nuclear attack, these agencies turned towards new forms of risk such as environmental hazards, failing states or (later in the 2000s) terrorist attacks. Consequently, "their mandates were then expanded from civil *defense* (protecting non-combatants against the effects of military attack) towards civil *protection* (protecting the population against a diffuse set of potential emergencies and disasters)" (Hagmann and Cavelty 2012: 83).

As a result contemporary security policy in the UK draws on an 'all hazards' approach, which is perfectly illustrated in the production of national risk registers that function as a complement to the NSS (Hagmann and Cavelty 2012: 84). In these comprehensive risk assessments all conceivable dangers and harms to the UK population are assessed according to their probability as well as potential impact.[1] This 'measuring of everything' comprises for example dangers through major flooding, severe space weather or cyber attacks. "The risk picture is likely to become increasingly diverse" (Cabinet Office 2010b: 19). As these assessments constitute the basis of governmental decisions on budgeting, political decisions on the distribution of financial and material resources are ultimately placed into the hands of a few security experts (Hagmann and Cavelty 2012: 90).

However, reading the reconceptualization of UK security discourse as a shift from geopolitics to biopolitics, or security to risk, would be misleading. Rather than the disappearance of a geopolitical security logic, we can see a complex re-combination of classical security and novel risk rationales and practices. Even a biopolitical approach to security, drawing on routinized assessment and management of anonymous risks, requires the construction of a collective identity – a valued referent object whose wellbeing it seeks to protect. Drawing on Bigo's (2006) genealogy of the concept of *protection* we can distinguish between different constructions of such valued referents in a national context. According to Bigo, the first notion of protection refers to the protection of the territory. It constructs a clear-cut inside/outside distinction. Drawing on a Machiavellian metaphor Bigo summarizes: "the prince is both the people and their government, and the idea is that sovereignty is harmed, if the borders are not protected" (Bigo 2006: 88). The second discourse turns the focus to the inside of the territory and is concerned with the vulnerability of some of its parts – by prioritizing and by assessing zones of high or low profile of risk. It is a "disciplinary moment which disaggregates the body of the prince by protecting some parts better than others" (Bigo 2006: 89). The third discourse, lastly, is concerned with the protection of the population and draws on a problematization of movement. Hence, protection cannot be located within a particular space or territory. "The focus needs to be changed by protecting people on the move" (Bigo 2006: 89).

Against the backdrop of this typology, the reconceptualization of UK security policy can be understood as a rearticulation of the concepts of defense, security and protection. The broadening of UK security thinking and the fusion of civil protection and security meant that not only the territory but also the individual, including the vulnerable amongst its people, were constructed as the subjects to be protected (Cabinet Office 2009: 17). At the same time, security policies even construct an obligation to protect people abroad – as is the case in the discourse on humanitarian interventions, the protection of refugees or climate change adaptation. Protecting the vulnerable, however, also always implies monitoring and controlling them for the sake of their wellbeing (Bigo 2006: 91). The result of this rearticulating of the protected referents is a conception of security policy in which we find very heterogeneous political technologies: from all-encompassing surveillance, monitoring and risk assessment, to measures to increase the preparedness of the vulnerable and rather traditional security measures such as border controls or military interventions. The overall effect is not a diminishing security logic but rather the total expansion of this logic. The paradigmatic expression of this totalization in the UK is the formulation of an 'all hazards' (Hagmann and Cavelty 2012: 84) approach to security – a security approach that seeks to be prepared for any conceivable danger, as demonstrated in this quote: "The National Security Strategy sets out [...] to ensure a secure and resilient UK by protecting our people, economy, infrastructure, territory and ways of life from all major risks that can affect us directly" (Cabinet Office 2010a: 9).

The narrative of the UK in an age of complexity

The 'all hazards approach' to security is informed by a political narrative about *the UK in an age of uncertainty* (Cabinet Office 2010b: 3). This meta-narrative, which constructs the UK's view of the world and its own position in it, draws on a particular self-image of the UK. This self-image is informed by the status of *the UK as an island* and *as a global hub*. It stems from the concept of the network and starts with the assumption that post-modern capitalism is characterized by the spread of (global) *networks*:

> The increasingly networked, interdependent and complex nature of modern society, and the critical systems which underpin daily life will, over the coming years, increase both the UK's vulnerability and the potential impact of civil emergencies. The interconnectedness means that a relatively small event, such as an electricity outage or loss of key information and communications networks, irrespective of the cause, can potentially lead to a cascade failure, with impacts on a wide range of critical services, such as water, transport and gas.
>
> (Cabinet Office 2009: 89)

Such networks provide the material and immaterial infrastructure for the global flow of goods, people, information etc. The UK, due to its status as an island which has historically been dependent on its external trade relations, and its history as a major imperial power, then understands itself as a *hub* in this complex grid of global networks: "The UK is a highly 'globalized' nation. This reflects our long history as an open, outward facing, trading nation and a hub of global activity" (Cabinet Office 2008: 49).

However, this self-defined position is not necessarily considered as an advantage. This is because social risks and economic bads are considered to be traveling through the same networks as economic goods: "In our interconnected world, the effect of violence in one region can spread to other more stable areas through refugee flows, terrorist activity, and organized crime groups, all of which can have an impact on our own security" (DfID *et al.* 2011a: 9). These dangers circulate through the very same channels and networks as societal goods, as in the case of trafficking, illegal immigration, diseases and computer viruses. Now, while each of these networks is already complex in itself, following the storyline of the *UK as a hub* they interact in complex ways: "Overall the risk profile continues to be diverse, with no single risk dominating. It also remains *complex* and *unpredictable*, with links *randomly* and *suddenly emerging* between previously *unconnected* events" (Cabinet Office 2009: 89). This fact makes calculation and probability assessment of cascading events impossible (Cabinet Office 2010b: 25). In the case of a complex emergency, the metaphor of the *UK as a global hub* turns into one of extreme vulnerability: "The UK's position as a hub of international communications, travel and migration means that it can act as a stage where international events can be played out domestically [...]. Domestic security cannot be separated from overseas security" (Cabinet Office 2009: 38). The meta-narrative of the UK in an age of uncertainty is generally introduced with the statement that there is a new "global landscape" and "this new global picture also presents complex, diverse and unpredictable challenges" (Cabinet Office 2009: 19). As the empirical examples show, this threat image is constructed along the lines of collective symbols borrowed from complex systems thinking including such as metaphors systems, networks and connectivity (Kaufmann 2013: 58) as well as biology with a focus on emergence, adaptation and organization (see also Chapter 4). The resulting *problematization of complexity* comes along with the conviction that stagnation or absolute stability in the contemporary world is not possible any more. On the one hand, therefore, modern government in general and security politics in particular require constant movement, recalibration and adaptation to changing conditions. On the other hand, however, movement or change represents a sort of danger in itself (Vaughan-Williams 2010: 1072). Given the two-fold framing of movement and change as opportunity and threat, the conception of security policies begins to change. Under the headline "stationary is dead," for example, the British think tank E3G states that "[p]reparing for the future means rejecting stationary as a guide to future outcomes. The first presumption should therefore be that all critical systems will be vulnerable without adaptive measures" (Mabey 2011: 103). Protective security policies must be as versatile as the perceived threats: "In an age of uncertainty, we need to be able to act quickly and effectively" (Cabinet Office 2010b: 5).

The perception of a necessity for adaptive flexibility resonates with a general skepticism against central control or steering: "The Government alone cannot protect people from all the consequences of natural disasters. Human and social resilience, often at the community level, will continue to be crucial to ensuring British citizens' future security and well-being" (Cabinet Office 2008: 42). While stability and balance have long been major aims in international security politics, they are not so any more (DfID *et al.* 2011a: 2). The stability concept, therefore, has thus been displaced by more flexible concepts such as *resilience*:

> we cannot prevent every risk as they are inherently unpredictable. To ensure we are able to recover quickly when risks turn into actual damage to our interests, we have to promote resilience, both locally and nationally.
>
> (Cabinet Office 2010b: 26)

The reason is that *resilience* draws on an epistemology of *becoming*, i.e. it assumes a continuous evolution, change and adaptation of the (natural or social) systems that it addresses. Experience drawn from the past and abstract knowledge drawn from statistical regularities become useless in governing a future which is marked by surprise and radical novelty (Aradau 2014: 7).[2] The ideal political subject of resilience is conceptualized as a continuously evolving system, which is able to cope with major disruptions because it quickly adapts to novel situations. "Resilience involves not only holding the line, but also making headway via the ameliorating force of increased fitness and vigor" (Evans and Reid 2013).[3] This shifting responsibility from the protector to the vulnerable can be understood as a partial shift from security as control, or defense, to a notion of security as resilience (Corry 2014). The notion of resilience was not brought into security discourses through a broadening of security – even in the 1970s it was already applied in strategic thinking on civil defense against complex nuclear threats (Duffield 2011: 11). But it was certainly rediscovered through this process and is heavily promoted as a core governmental principle through the storyline of the *UK in an age of uncertainty*.

Central themes and nodal points in UK climate security discourse

In Chapter four I described the process of a securitization of climate change in the UK. I could show how more and more actors from different subject positions took up climate change and security storylines. By the end of 2007 there was a UK discourse coalition revolving around climate security storylines that comprised major political actors such as the then foreign secretary Margaret Beckett and premier Tony Blair, security related think tanks, such as RUSI or E3G, peacekeeping NGOs such as International Alert, humanitarian and environmental NGOs such as Green Peace or Christian Aid as well as bureaucrats from the MoD, FCO and DfID. It is hence fair to say that a climate security discourse reached a high degree of *discourse structuration* in the UK (Interview with a UK climate security analyst, London October 21, 2011). In 2008 the first NSS was published by the UK government, which for the first time set out the comprehensive security approach outlined in the previous section. Hence, the question of whether there is a causal relationship between these two developments – the securitization of climate change and a broadening of the general security approach of the UK – arises. To provide an answer to this question, it is first important to present the main storylines articulated by a UK discourse coalition on climate security.

Climate change risks and national security

To study the impact of a securitization of climate change in a broader UK security field (or vice versa), it is insightful to look at how climate change is represented and treated in major UK security documents. In a nutshell, one can say that climate change in the UK has been securitized on the basis of a risk and resilience rationality. Clearly, there is not one single reason for the prominence of resilience in UK climate security, but a crucial factor was the rejection of simple models of climate conflict by large parts of the development as well as the foreign policy community. Climate security arguments were rejected due to their simple, linear and mechanistic correlations drawn between climate change and violent conflict. The discourse of resilience rejects such simple notions of causality and replaces them with a focus on emergent complexity (Chandler 2014: 40). Stressing the complexity in linkages between climate change and security, with a focus on a context-sensitive, locally specific and multi-causal examination of such linkages (Vivekananda *et al.* 2014: 492–494), then allowed actors from the development field to buy into the discourse again.

The overall discursive structure distinguishes between *two channels* through which climate change becomes security relevant to the UK. The first is in a direct manner by producing or amplifying a range of different dangers for the UK territory, infrastructure or population that are assessed through risk technologies. Such an approach is completely in accord with risk management practices in other security-related domains:

> Because of the scale and speed of the risk they pose, those phenomena [climate change or global pandemics, author's note] have similar

> potential to other security challenges to threaten our normal way of life across significant areas of the country with little warning. Moreover, our approach to them [...] is similar to our approach to other national security challenges, including terrorism.
>
> (UK Cabinet Office 2008: 15)

To cope with the direct risks posed by climate change, the UK has implemented a comprehensive climate impacts program including a periodically conducted national climate risk assessment. The first report from 2012 comprises an assessment of over 700 different climate-related risks for the UK within the timeframes of 2020, 2050 and 2080 (Defra 2012: viii). The risk assessment provides the knowledge resources for the development of a National Adaptation Programme (NAP). As the executive summary of the report states:

> Although we do not know the likelihood of future changes in the UK's climate, we know enough to present a range of possible outcomes, which can be used to inform adaptation planning. For this purpose potential climate risks to the UK have been categorised according to their magnitude, 'confidence' and the 'urgency for action.'
>
> (Defra 2012: xi)

Resilience here is not so much a discourse of unknowability, as suggested by some authors, but resilience is explicitly related to practices of climate risk calculation and the compiling of comprehensive risk registers. As Dunn-Cavelty *et al.* note resilience is "interwoven with other forms of security governance including technologies of risk, insurance, and preparedness" (Cavelty *et al.* 2015: 6). The close relation between prediction and resilience is perfectly mirrored in the foreword to the 2013 NAP Report *Making the country resilient to a changing climate.* In this foreword, the Secretary of State for Environment Owen Paterson argues that:

> Britain's expertise in areas such as weather forecasting, flood modelling, infrastructure and insurance are already coming to the fore to prepare us for the kinds of events we might see more often. Indeed, the UK is already one of the global leaders in this industry of the future.
>
> (Defra 2013: 1)

Climate change is the driving force behind the emergence of a genuine 'industry of the future' – a private sector specialized in risk anticipation and assessment, which will deliver its services to the whole defense sector.

A second risk framing of climate change is concerned not so much with the direct impacts of climate change but with the "consequences of consequences" (Government Office for Science 2011b: 7). This framing constructs climate change not as a risk in itself but as a major driver of insecurity (Cabinet Office 2009: 32). You can find this representation of climate change in all major security documents of the UK. Illustrative is the representation of climate change as a *ring road issue*. A ring road issue is understood as "a driver that is so pervasive in nature and influence that it will affect the life of everyone on the planet over the next 30 years" (DCDC 2010: 6). Climate change is not only a master threat but also perceived as a complex system whose secondary impacts are not predictable through traditional risk assessment (Interview with a UK climate security analyst, London October 21, 2011). The climate security discourse hence resonates very well with the overall storyline of the UK in an age of uncertainty. This is expressed in the following example:

> the impact of technology, migration and trade has led to a highly interconnected globalised world and it will be difficult for the UK to isolate itself from the global economic and geopolitical shocks that look certain to be experienced in a warming world.
>
> (Government Office for Science 2011b: 118)

What is striking here is the kind of epistemological interplay that characterizes both climate research and security thinking. On the one hand, climate system research has informed such notions of complexity which led to a rethinking in contemporary security policy (Kavalski 2011: 6–7; Mayer 2012: 171–174; Price-Smith 2009). On the other hand, climate change-related threats are themselves understood as a major form of such complex and rapid security risks: "our better understanding of the imminence and impact of climate change [...] demonstrate[s] the interconnectedness of global security challenges in the twenty first century" (Cabinet Office 2009: 25).

Climate change and a demilitarization of security

The second characteristic of the national discourse on climate change and security is that it seemingly promotes a demilitarization of security thinking – an observation which stands in contrast to the expressed fears that a securitization of climate change could lead to a militarization of climate politics. In the UK national discourse on climate change and security, military operations are not at all articulated as a possible solution to climate change-induced security threats (Phone interview with a member of the UK MOD, September 18, 2012). Rather, the dominant discourse articulates the growing demands for (humanitarian) military engagement as a part of the threat scenario (Interview with a UK climate security analyst, London October 21, 2011).[4] If climate change accelerates in an unmitigated manner, it is anticipated that the UK military will see itself forced into a number of humanitarian relief and crisis interventions: "The

UK, as an active member of global governance structures, faces more calls to help formulate and contribute to international interventions in regions where tensions arise or are exacerbated by climate change impacts" (Government Office for Science 2011b: 44). A military response is not at all articulated as an appropriate response to the danger of climate change. Quite the contrary, it is being assumed that climate change could put serious constraints on the UK's future defense capabilities. On the one hand, this is because climate change poses a threat to the 'defence estate' (MoD 2010c: 11); "[t]raditional war-fighting capabilities will [...] continue to be needed, but with the prospect of being deployed under more challenging operational conditions which would include higher temperatures and water scarcity" (Government Office for Science 2011b: 46). On the other hand, growing calls for stability and humanitarian missions could *overstretch* the military capacity of the UK and hence undermine the capacity to deter or to defend itself against other military threats (MoD 2010a: 14).

The fact that military measures are not articulated as an appropriate response to deal with climate change, however, does not mean that such a climate security framing would be unproblematic. By projecting a catastrophic climate change in the future, the requirement of such interventions in the future appears as a mere *matter of fact*, such as in the following example:

> More frequent and severe extreme weather impacts will cause urgent humanitarian emergencies that could accelerate collapse and conflict in vulnerable states. Thus, security in the 21st century will require a major increase in the capacity to launch coordinated international humanitarian and preventive missions.
>
> (Mabey 2011: 133)

This implies a *naturalization* of a politics of intervention, which takes the desirability and feasibility of such policies out of public dispute and hence clearly represents a form of *depoliticization*. Furthermore, such an anticipation of a dangerous future constitutes the basis for the allocation of budgets in the present.

A similar rationale – of an apparent demilitarization of security thinking – informs the UK discourse on CIM. While all studied sources articulate CIM as a major concern for the UK, they do not call for a repressive management of this problem, e.g. through stronger border protection or policing measures. The authoritative Foresight report on *Migration and Global Environmental Change* (Government Office for Science 2011a) is a case in point. The report, which received major international attention, not only rejects a repressive treatment of CIM but even frames that as a possible risk: "Preventing or constraining migration is not a 'no risk' option. Doing so will lead to increased impoverishment, displacement and irregular migration in many settings" (Government Office for Science 2011a: 9). On the contrary, the authors of the report suggest understanding CIM as a form of a "'transformational adaptation' to environmental change" (Government Office for Science 2011a: 7). "In particular, planned and facilitated approaches to human migration can ease people out of situations of vulnerability" (Government Office for Science 2011a: 10).

Following a logic of resilience, CIM is rearticulated from a security threat into a measure to reduce climate change security implications by promoting long-term resilience (Bettini 2014: 189; Methmann and Oels 2015: 58–59). Yet again, one has to note that such a non-military take on CIM is not unproblematic. On the one hand, fostering migration as a proactive adaptation measure implies that authoritative decisions are made in the name of the vulnerable, which is particularly evident in the case of planned resettlements. On the other hand, such a proactive approach to manage CIM is problematic as it means that decisions over people's lives are made on the basis of insecure anticipations of the future. It "presupposes that climate change is an unavoidable reality and fact that needs to be lives with" (Methmann and Oels 2015: 63).

Climate resilience as a governmentality of sacrifice

Studies of climate change discourses from a critical security perspective have so far focused *either* on the construction of climate change in either terms of existential security and exceptional measures *or* in terms of risk and resilience (Corry 2014). There is as yet no study, however, that explains the relation between these two obvious features – exceptional threat construction and risk-based self-government – of climate change discourses (Methmann and Rothe 2012).

The case of the UK, however, shows that the discursive construction of climate change as a major external threat and an unfolding global catastrophe construct two forms of subjectivity that seek to activate the self-government potential of individuals. The first is the *ecological citizen*; the second is the "politically debased *subject of resilience*" (Reid 2012, emphasis added). First, I would argue that the sustainable or environmentally responsible subject (Luke 1999: 136) that governs its own environmental performance, for example by budgeting its personal carbon output (Paterson and Stripple 2013: 345–347), becomes linked to the climate security discourse through a *storyline of sacrifice*. Usually, governmentality – or pastoral power as its generative model – is presented as a form of power that promotes self-government to increase the wellbeing of the governed subjects (Foucault 1988). However, governmentality can also mean to promote self-government to increase the wellbeing of the *political community* or collective body (Foucault 2007: 129). A *governmentality of sacrifice* requires subjects to live healthily, to circumvent dangerous places or situations, etc. for the sake of the collective, as expressed in the following example: "Everyone has a role to play in building the UK's resilience" (Cabinet Office 2009: 31). A *governmentality of sacrifice* requires the construction of a common identity through a logic of equivalence, i.e. of a political community threatened from an outside enemy/danger.

An example that I find highly illustrative of this close marriage between forms of environmental self-management and the construction of existential macro-threats are political posters in World War II (WWII) that aimed at mobilizing the self-governmental potential of UK citizens. Such posters, which are displayed at the Imperial War Museum in London but unfortunately could not be reproduced here for copyright reasons, show that the demands that are commonly directed towards environmental subjects today – sustainable consumption, energy saving behavior – were also demanded in the crisis situation of WWII. UK citizens were asked if their "journey is really necessary"?; they were told to grow their own vegetables and not to waste food. This is not to empirically compare the phenomenon of climate change to the threat of Nazi Germany in WWII. The important thing for my argument is that both phenomena are constructed as totally external and existential enemies/threats and both are linked to conservationist demands underscored with a moral imperative to "do your bit." The discursive storyline that links an overarching threat to personal responsibility is highly interesting for our case:

> Individuals have an essential role to play in national security. We can all contribute, for example by being vigilant against terrorism, and by planning for, and taking a more active role in responding to, civil emergencies [...]. We can also play our part in tackling the longer-term challenges, such as climate change.
>
> (Cabinet Office 2008: 59)

Recall that, as in the case of Nazi Germany in the WWII posters, global climate change is constructed as a total and external enemy: "Climate change is potentially the greatest challenge to global stability and security, and therefore to national security" (Cabinet Office 2008: 18). This construction of a major external threat, then, is linked to calls for individuals to do their bit. As exemplified in the posters, this demand is not enforced coercively but works through the subject's conscience. Following the work of Paterson and Stripple (Paterson and Stripple 2013), I would argue that climate governance in the UK is informed by a particular carbon governmentality, which obliges individual actors including citizens, communities, or businesses to carry the burden of climate mitigation through different technologies of the self.[5] Now we are able to see how such a governmentality of *individual sacrifice* is legitimized through the discursive construction of climate change as an external and total threat and the similar construction of legitimate political community – as a referent. While Germany in

WWII, however, clearly represented an external and existential threat to the UK, climate change represents an external enemy only in that it is discursively externalized, anthropomorphized and thereby isolated from its proper socio-economic context. The individual citizen then has to pay for, for example, the bad performance of a fossil fuels-based political economy.

Not only in mitigation policies but also in the field of adaptation, climate security storylines are linked to a mode of subjectivity – the resilient subject – that requires continuous self-regulation (Chandler 2014: 56–57). As outlined above, both in national adaptation and disaster risk management programs and in international adaptation support, the UK builds on the concept of resilience. The concept of resilience requires subjects to accept both the complexity of the world and its proneness to crisis (Reid 2012: 74). The apocalyptic nature of the threat as an unfolding but incalculable catastrophe requires being in a constant status of preparedness: "It's all about preparation for the unknown [...]. It's about being prepared, and not just in a military sense" (Phone interview with a UK climate security analyst, December 14, 2011). This requires political subjects to take up responsibility for their own well-being: "Our people themselves, through voluntary organizations, civic society, private industrial sector and all levels of government need to work together to tackle domestic threats and develop our resilience" (Cabinet Office 2009: 29). Resilience is a form of self-government – a form of taking-care-of-oneself – that works through the invocation of anxiety and which replaces the political behavior of subjects with an adaptive one (Evans and Reid 2013: 364).

Furthermore, resilient subjects have to continuously adapt to changing conditions, which in developing countries, for example, could mean changing their lifestyles, means of subsistence or even their residence. As it has been argued, all of this has to be done for the sake of the collective system. This can be the nation, such as in the case of the UK national resilience program, or the community, as in the case of community-based adaptation support promoted by the UK (Cabinet Office 2009: 31). Subjects live in a permanent state of exception as the UK is threatened not by a present and concrete enemy such as Nazi Germany but by a range of haunting, non-sizable risks. The dominant storyline that stands behind this permanent threat scenario is that of *the UK in an age of uncertainty*. A threat that is so overarching and at the same time so incalculable or complex such as climate change can only be fought through the everyday practices of everybody (Methmann and Rothe 2013).

Securitization as climate mainstreaming

To study the forms of institutionalization and the policy implications of climate security discourse in the UK, it is necessary to turn the focus from articulations of climate change and security to the practices of governmental actors and the question of how these have adopted and approached climate security. Not only has the UK established an explicit Climate and Energy Department, all security-relevant ministries – the FCO, the MoD and DfID – have also established departments or working groups on climate change and security. In 2010 the government even appointed Rear Admiral Neil Morisetti as the United Kingdom's Climate and Energy Security Envoy. And John Ashton, an environmental security analyst who has been a prominent voice in UK climate security discourse, was appointed as the Foreign Secretary's Special Representative for Climate Change.

By taking a closer look at how different ministerial actors have approached climate security one can identify a particular logic of *climate mainstreaming* (Methmann 2014: 5). This term refers to the phenomenon that different actors from the development, security and foreign policy fields that previously did not deal with climate change adopted the issue and brought it in line with their own rationalities, routines and practices (Phone interview with a member of the UK FCO, December 15, 2011). A good example for such climate security mainstreaming is how the issue was approached by DfID and other actors in the field of international development policy.

Climate change and international development

In Chapter 4 on the conceptual roots of climate security discourse I have outlined how in the 2000s one could see the evolution of a development–security nexus – a growing interplay between discourses and practices of development and security policy. In an interlinked world, instability and turmoil as a result of underdevelopment abroad were increasingly seen as a security risk for the country. As a result actors from the defense or peacebuilding field were increasingly integrated in UK international development policy. In the late 2000s, climate change was added as a third ingredient to this nexus, which thus became a *climate–development–security nexus*. You can imagine this climate–development–security nexus as an interdiscursive field which developed a common discursive vocabulary for all three specialized discourses, and through which concepts and related practices could travel between the discourses. The existence of this interdiscursive field explains why, for example, concepts such as vulnerability, risk, sustainability and resilience have become prominent in all three fields: climate, development and security policy (Reid 2012: 72–73).

The 2011 Foresight report on the international dimensions of climate change is illustrative of the general rationale behind this climate–development–security trio. It states that many of the harmful consequences of climate change will materialize in the developing world. Yet drawing on the self-understanding of the UK as a global hub, the global connectedness of the UK means that it cannot ensure its own immunity to these negative effects. This makes the vulnerability of communities in the global South an immediate concern for the UK. Hence, "A recent review of how the UK government should respond to humanitarian disasters has recommended that building resilience to the threat of climate change in countries at risk is integrated into the core DfID programme" (Government Office for Science 2011b: 45).

At the same time, NGOs active in the field of peacebuilding started reframing their work in relation to climate change. Peacebuilding was increasingly rearticulated as both a climate adaptation as well as a conflict prevention measure, as this quote demonstrates:

> A society that can develop adaptive strategies for climate change in this way is well equipped to avoid armed conflict. And a society that can manage conflicts and major disagreements over serious issues without a high risk of violence is well equipped to adapt successfully to the challenge of climate change.
>
> (Smith and Vivekananda 2007: 32)

Harris (2012) comes to the conclusion that the securitization of climate change in the UK so far did not find its expression in UK development policy. She substantiates this claim by looking at the concrete projects that were funded through UK ODA as well as the UK *International Climate Change Fund* (ICF).[6] Yet, I would argue that the climate change and security discourse has affected UK development policy at a more fundamental level: funding adaptation and mitigation in the global South was increasingly considered a part of a broader security strategy (Phone interview with a UK climate security researcher, November 2, 2012). It was precisely this rationale that stood behind the establishment of the ICF: "Taking action on climate change is in Britain's interest. We cannot have food security, water security, energy security – or any form of national security – without climate security" (DfID *et al.* 2011b: 2). The national security strategy update 2009 explicitly mentioned the ICF and the funding of resilience projects as a measure to build capacities and stability overseas (Cabinet Office 2009: 53). These examples illustrate how the climate security discourse managed to integrate security with economic rationales.

Indeed, the climate change programs in the UK's development policies basically take the form of investment practices – as financing mechanisms for adaptation and mitigation projects (DfID *et al.* 2011b: 6–9).[7] The security-based storylines do not oppose a market-based or economic approach. On the contrary, security articulations easily translate into an economic discourse. I would argue that this goes mainly back to the resilience paradigm, which dominates the UK's security policy. The concept of resilience is "abstract and malleable enough to encompass the worlds of high finance, defense and urban infrastructure within a single analytic" (Walker and Cooper 2011: 144). Risk and resilience provide coherence to economic and national security rationalities that would not easily fit together in a traditional sense of security. Under a resilience paradigm it makes perfect sense that communities must *invest* in their own adaptive capacity and that development policy should finance disaster insurance schemes. The nexus between climate change, fragility and conflict is counter-posed with a positive model linking climate change, resilience and peace (Vivekananda *et al.* 2014: 493).

What is more, DfID incorporated climate change into its broader strategic priority on "building stability overseas," for example as one crucial factor in conflict early warning and horizon scanning practices:

> We will establish an Early Warning System that will take a global view of countries in which political, economic and security shocks over the next 12 months could trigger violence. This will be underpinned by all-source analysis from the Cabinet Office, FCO, MOD and external experts.
>
> (DfID *et al.* 2011a: 20)

DFID's strategic focus on stability abroad has had clear cut political implications, such as the establishment of a stabilization unit and so-called stabilization response teams as well as military and civil forces that could intervene in humanitarian and crisis situations (DfID *et al.* 2011a: 19).

The establishment of these stabilization teams is not part of the UK's climate governance, nor is it a direct effect of the securitization of climate change. Nevertheless, the international impacts of climate change have become the most important justification for why it is important for the UK to provide stability abroad (Phone interview with a UK climate security researcher, December 13, 2011). It is not the securitization of climate change alone but its integration into a climate–development–security nexus, which might justify and fuel a growing role for the UK military in global crisis situations:

> Protecting civilians is at the core of the UK's policies to prevent, manage and resolve conflict. The UK Government believes that humanitarian space needs to be protected and expanded, including in fragile and conflict-affected states.
>
> (DfID *et al.* 2011a: 3)

And as outlined above protection always implies some degree of intervention into the affairs of the protected as well as some degree of control.

Climate change, active diplomacy and foreign policy

Climate change was not only mainstreamed into UK development policy; the way climate change was integrated into UK foreign policy also followed a climate mainstreaming logic. The FCO quite early on adopted climate change as a central topic of its work. Since 2000, the FCO has had an environmental policy unit dealing with climate change. In 2004 it installed an Energy and Climate Change Unit. Since 2009 there has been a distinguished group within the FCO dealing with climate change and security. The way that the organization approached the issue was through a mainstreaming strategy: already existing projects, e.g. on conflict prevention, were reframed as climate change-related projects. However, the FCO did not construct climate change in neutral terms but through the prism of a climate change and security storyline (Boas 2015: 86–93).

I would argue that the articulations of climate change and security storylines through the FCO were embedded in the broader UK security discourse outlined above. Of particular interest, in this respect, is the self-understanding of *the UK in an age of uncertainty*. If one acknowledges this narrative, the FCO's mainstreaming of climate change through the lens of foreign policy – as part of a broader *active diplomacy* approach (FCO 2006) – can be regarded as a form of *preemptive security policy*. Following Anderson, preemption refers to technologies of security governance that aim at actively modeling or changing a future reality in order to facilitate desired developments and prevent harmful ones (Anderson 2010: 789–790). Climate change mitigation policies – as well as adaptation in hard-hit regions and countries – are exactly such attempts to create a future that is, as much as possible, in favor of UK defense interests.

The FCO's 2006 report on the international priorities of the UK, for example, states: "What happens abroad has never mattered more for our security and prosperity. In an age of rapid global change, the task for Government is to seek to understand and influence the world for the benefit of our people" (Blair in FCO 2006: 3). Climate change here represents one master variable or security driver in the future alongside other global trends such as globalization and rising economic complexity, population growth and migration, resource depletion and growing energy demand (FCO 2006: 10–7). As a result of the complex interaction of these trends, any of the imagined possible futures will come into being.

The UK's understanding of itself as a global hub stresses the interdependence of the UK in a complex world. Climate change and security discourse, then, reinforces this self-perception: although the UK regards itself as a pioneer in climate protection efforts, it cannot achieve climate security on its own (Phone interview with a UK climate security researcher, December 13, 2011). This dilemma for the UK leads to the concept of an *active* or what I would call *preemptive diplomacy*. Conventionally, diplomacy is regarded as the art and praxis of *maintaining* something, namely good relations between states, or *managing/balancing* something, for example diverging interests between states (Oxford Dictionary Online 2015). An active diplomacy is different in this respect – it does not seek to maintain but to *produce* a desired international order or to *prevent* a range of undesired orders in the future.[8] The FCO's concept of *climate diplomacy*, then, can be seen as a crucial part of such an active diplomacy concept. As such, it represents a part of a security politics that tries to make the deployment of defense measures in the future unnecessary by actively shaping the future: "it represents something a lot bigger than climate change. It's a new avenue for international relations" (Interview with a UK climate security analyst, London October 20, 2011). The 2010 Strategic Defense Review, for example, expresses this approach "to shape a stable world, by acting to reduce the likelihood of risks affecting the UK or our interests overseas, and applying our instruments of power and influence to shape the global environment and tackle potential risks at source" (Cabinet Office 2010a: 9).

This preemptive or active climate diplomacy builds upon the two pillars of mitigation and adaptation:

> By taking an active role in particular aspects of mitigation and adaptation to global climate change, the UK can help other countries gain from its expertise and experience, and so influence international responses and organisations in a way that supports UK approaches.
>
> (Government Office for Science 2011b: 92)

With respect to mitigation, the UK pushes other countries to take up more active emission reduction commitments and policies. Against the backdrop of the climate change and security discourse, it is in the UK's national interest to keep climate change to a tolerable level – and this requires the cooperation of a large number of countries. For this reason, the FCO decided to bring the issue to the UNSC when the UK had the presidency in 2007 (Boas 2015: 90). And hence, the UK uses its international relations to "shape international policy and direction" (Government Office for Science 2011b: 92). The second pillar focuses on adaptation. Here, the FCO uses its diplomatic activities and contacts to raise the awareness of developing and emerging countries (Boas 2014). This also implies interfering in the sovereign affairs of other states:

> in many countries that face the double-headed problem [of climate change and violent conflict, author's note], the government is going to be either unwilling or unable – or both – to take on the task of adaptation and peacebuilding. [...] The task of helping communities adapt to climate change cannot be left to such governments.
>
> (Government Office for Science 2011b: 23)

The strategy of climate diplomacy outlined in this section, it is hoped, would then enable countries to increase the preparedness of their populations, which in turn would prevent them from becoming dangerous in the future.

Climate change and defense – greening the military

Above I argued that a military-centered security policy was not articulated as a solution to climate change-related security threats but rather as part of the threat scenario. The MoD's appropriation of climate change and security storylines mirrors such a framing. Rather than the adoption of exceptional security measures, in climate governance one can notice that defense actors increasingly integrate ecological reasoning into their operational routines (Phone interview with a UK climate security researcher, November 9, 2011). Climate change is mainstreamed into defense policy – a process that is sometimes summarized under the label of *greening the military* (Interview with a climate security analyst, London, October 20, 2011).

First and foremost this means that the MoD adopted the empty signifier of sustainable development[9] as its primary climate policy concept. The concept of sustainability is then reinterpreted in the defense context and rearticulated as "defence sustainability":

> SD [sustainable development, author's note] in Defence is more specifically defined by two Defence SD Principles: Principle 1: Defence must be resilient to current and future environmental, social and economic threats (adaptation). Principle 2: Defence must realise the positive and minimise the negative impacts that Defence activities can have on the environment, people and the economy in the UK and overseas (mitigation).
>
> (MoD 2011: 1)

Beginning with the field of adaptation, one can notice that the defense sector adopted a risk approach to enhance the resilience of defense estate and infrastructure. The aim of this risk management has been to preserve the UK's defense capabilities by adapting to: "the risks to global security presented by the complex geopolitical interactions resulting from a changing climate, as well as the risk to our own establishments and equipment from the impacts of climate change" (MoD 2010b: 2). For the protection of the defense infrastructure and estate, the MoD developed a Climate Impact Risk Assessment Method (CIRAM). This is used by commanders of different bases to assess the vulnerabilities and capabilities of their site. Although the climate risk assessment was prompted by the central government and was conducted by all departments,[10] the MoD represents a special case because it tied in with the MoD's own routines: "CIRAM is a risk assessment methodology based on MoD's own risk management methodology" (MoD 2010d: 3). The result of this assessment was the production of a "Climate Resilience Risk Register" (MoD 2010d: 5). What becomes apparent is that the defense ministry's approach in dealing with climate change threats equals the risk-based 'all hazards' approach that characterizes current UK security policy in general.

In the case of mitigation, the MoD articulates a twofold global challenge of climate and energy security. Following this framing, the British military is vulnerable to both climate risks and to energy supply disruptions and price fluctuation:

> MOD is a substantial user of energy, both from the grid and from the use of fossil fuels, which exposes us to price spikes and supply side risks. As part of the MOD response to climate change, an opportunity presents itself for the Department to become less dependent on these fuels and therefore less exposed to significant price spikes.
>
> (MoD 2010a: 6)

The concept of sustainability transforms the threat scenario of a climate and energy challenge into a win–win metaphor. Sustainability is rearticulated as both the sustainability of earth systems and natural resources and the sustainability of defense operations. A major focus of mitigation efforts by the MoD and the military is hence on low carbon *technologies*. According to the defense sustainability storyline, the military could even take up a central role in the development and distribution of such technologies, as the defense sector is not subject to the restrictions of the free market:

> This [the adoption of low-carbon technologies, author's note] should be achieved by exploiting emerging civilian applications and technologies wherever possible, but also driving, through military research and development, operationally beneficial low-carbon technologies where they are not developing rapidly enough commercially.
>
> (MoD 2010c: 7)

Hence, a large part of the defense sector's mitigation activities revolve around energy use, energy efficiency and the substitution of fossil fuels (MoD 2010b: 5–6). It perfectly fits with broader climate change mitigation discourses that focus on a technologically driven shift towards a low carbon economy. Also in other attempts to contribute to national climate mitigation efforts British defense actors basically imitate the dominant *carbon governmentality* (Paterson and Stripple 2013). For example, in order to reduce emissions from their military supply, the MoD has obliged enterprises in their supply chain to participate in the Carbon Disclosure Project (CDP) (MoD 2010b: 8).[11]

What one can see from the way the British MoD is integrating climate change into its operations is that even here the operationalization does not follow a logic of exceptionality, i.e. it does not break with the established rules. Quite the contrary: the MoD draws on a risk management approach that is in line with both the UK strategy on climate change in general and with its established risk assessment methods. To sum up, the MoD adopts climate change as an *already securitized* issue, but this securitization is not one of traditional security but rather one that follows a logic of risk and resilience.

A climate change and security assemblage

Due to the mainstreaming logic that has underpinned climate security in the UK until today, there is no overall coherent UK climate change and security policy that would follow a broader governmental logic. Given the diversity of climate security discourse which was revealed in this book and due to the complexity and non-linearity which is attributed to the threat of climate change, there simply cannot be one overarching logic in governing climate insecurity (MoD 2010a: 7). Nevertheless, I would argue that the different forms of institutionalization are not completely independent from each other. The notion of the assemblage borrowed from Gilles Deleuze and Felix Guattari (Deleuze and Guattari 1987: 504–505) provides the possibility of studying how the different forms of governing climate insecurities are assembled into a broader whole. The term 'assemblage' refers to formations of very diverse elements that are held together "without ceasing to be heterogeneous" (Allen 2011: 151). What holds the assemblage together is only the vague and empty signifier of climate resilience, which, in the UK since the late 2000s, more and more emerged as a consensual demand (Interview with a UK climate security analyst, London October 21, 2011). Resilience holds together the diverse practices of climate security not despite its vagueness but due to its creative ambiguity. It is precisely because the term is so malleable that different political projects are able to use it in different ways (Walker and Cooper 2011: 144) that it could become so dominant in UK climate security discourse.

Within this diversity of practices in the climate security assemblage we can distinguish broadly three different parts, each of which addresses a different aspect of the climate security field: *anticipating threats, producing resilience and forming a stable environment for the UK* (Table 7.1). These three parts of the assemblage link storylines, actions, objects and people horizontally, i.e. by linking programs from different departments, as well as vertically from the government level to the individual.

The first part of a climate security assemblage refers to the ways in which different actors work together to render climate change-related threats visible or calculable. Following this perspective, the establishment of a discourse coalition that produces different representations of climate security is in itself a particular political implication. In the assessment of climate change-related risks the climate security discourse coalition brings together the practices and articulations of very different actors such as governmental scientists including the Government Office of Science (Government Office for Science 2011a, b) or MoD's Development, Concepts and Doctrines Center (DCDC 2014), with defense or government-related think tanks such as the Oxford Research Group, RUSI or Chatham House.

Table 7.1 Assembling climate change and security practices in the UK

	Science–military	*Adaptation and resilience*	*International dimensions*
Actors/subjectivities	Climate security analysts, researchers, think tank members, defense strategists and analysts	Defense planners, civil protection agencies, resilient citizens, businesses and communities.	Foreign Office, UK diplomats, DfID, development and peacekeeping NGOs
Technologies/practices	Probability-based risk assessment and calculation, Cost-benefit analysis, Scenario techniques, Cartography	Investment in adaptation, governing through the community and self-government	Active diplomacy, awareness raising, governing through the community, monitoring and assessment of the vulnerable, managing climate-induced migration
Objects/materials	Reports, risk registers, risk matrices, statistics, maps.	From genetically modified crops to new insurance products such as catastrophe bonds, and resilient infrastructure	ODA, International Climate Fund, negotiation chambers, international laws and conventions
Binding element	A future made present	The preparedness of the people	Shaping a desired UK environment

Their cooperation and interaction through workshops and research collaborations is behind the growing convergence of methodologies and epistemologies in climate change sciences and defense or strategic thinking. The interaction delivers the knowledge required to produce climate resilience in terms of a calculation, visibilization, monetarization and imagination of future risks. Storylines and practices of a future climate become inscribed into objects such as reports, risk registers, risk matrices and statistics.

Second, adaptation and resilience bring together various governmental or departmental programs to form a comprehensive and integrated climate risk management. Climate risk management is integrated within a broader approach of *civil protection* that tears down departmental dividing lines. This part of the assemblage comprises, for example, the civil contingencies secretariat and related governmental programs on 'emergency preparedness,' 'infrastructure resilience' and 'community resilience.' Drawing on a 'multi-level, multi sector, bottom-up approach to resilience planning' (Cabinet Office 2009: 90), these programs incorporate businesses, communities and individuals into governmental climate risk management. One example of how this works is the establishment of Local Resilience Forums, in which local citizens and stakeholders meet regularly and exchange their actions and knowledge with regional administrations and the civil contingencies secretariat (Adey and Anderson 2012: 108). Emergency exercises at the local level are based on future scenarios such as those developed in the DCDC report on Global Strategic Trends in 2045 (DCDC 2014: 1–33). Such national exercises are a mode of directly enacting and embodying a dangerous climate future – such as a perfect storm event or major flooding (Adey and Anderson 2012: 102). The adaptation and resilience part of the assemblage, to sum up, links the practices of very different actors – from disaster planners, to private sector consultancies, resilient subjects, businesses and communities. It constitutes relations between hitherto unconnected elements and circulates tasks and responsibilities among these elements. Moreover, it continuously collects data brought together at the Cabinet level. In addition, different materials such as resilient infrastructure, insurance products or climate-resilient crops are assembled into UK adaptation governance (DfID *et al.* 2011b: 12; Government Office for Science 2011b: 93).

A last part of the climate security assemblage revolves around the international dimension of climate change. As argued by the important Foresight report on the international dimensions of climate change:

> To address the risks to the UK from climate change impacts overseas, it is crucial that government departments work across existing boundaries [...]. The impacts of climate change will need to be factored in across all areas of government policy which have an international dimension.
>
> (Government Office for Science 2011b: 120)

Cooperation between different actors here follows the aim of preventing dangerous implications for the UK. Climate diplomacy activities link the UK's efforts to other governments and try to foster a more integrated approach to mitigate climate-related threats at the regional level of the EU or the international level. At the same time, different fora such as the Security Council and international conferences are used to spread the debate on climate security linkages and to raise awareness in most affected countries. International conventions such as the UNFCCC and international law on CIM (Government Office for Science 2011a: 152–153) are actively addressed by the UK to reduce its vulnerability to the "consequences of consequences" (Smith and Vivekananda 2007: 7). Financial flows, e.g. through UK ODA or the ICF, become related to the prevention of climate risks. Even vulnerable communities in developing countries far away in this way take part in the production of climate security in the UK. On the one hand, local stakeholders and communities are empowered – at least, attempts to this end are made – and thus incorporated into the UK's governmental approach in dealing with climate insecurities. On the other hand, these risk groups are subject to monitoring and assessment, through UK scientific institutions such as the Government Office of Science but, of course, also through the work of developmental agencies or NGOs.

To sum up, due to the mainstreaming logic just outlined, climate change and security discourse bring together very heterogeneous actors, practices, storylines and objects without dissolving their peculiarity, i.e. different parts of the assemblage are not subjected to a broader governmental logic. In a *climate security assemblage* different elements are brought together not despite their differences but because of their difference.

Notes

1 This equals a *classic* approach of risk assessment, for example as used in the insurance industry (Rothe 2011: 332–333).

2 When the concept of resilience arose in the realm of the natural sciences in the 1970s, it was part of a broader critique of a Newtonian physics and its notion of inertia, linear movement and equilibrium (Duffield 2011: 11).

3 Just as with the construction of complex security threats, the resilience concept also draws heavily on motional metaphors stressing speed and non-linearity as well as biological metaphors stressing the emergent properties of risks. This stems not least from the fact that it was introduced to ecology and social sciences as a metaphor from the natural sciences where it denoted "the elastic ability of materials to return to their original form after being bent, compressed, or stretched" (Kavalski 2009: 539).

4 Studying the treatment of climate change in national defense documents, Brzoska (2012) comes to a similar finding: the dominant way in which the military treats climate change is as a growing area of activity for disaster relief and management operations.

5 For example, with the 2008 Climate Change Act the UK was the first country to impose on itself a legally binding carbon emissions reduction of 34 percent by 2020 and 80 percent by 2050 compared to the 1990 baseline. However, crucial

to the Act was that it did not prescribe any concrete policy instruments to mitigate climate change but institutionalized a comprehensive framework of carbon budgeting and calculation. For example, the Climate Change Act introduced a system of departmental carbon budgets controlled by a comprehensive system of audit and measurement (DECC 2008).

6 The ICF was established in 2010 and today represents the major British funding instrument for adaptation and mitigation projects overseas.

7 The ICF, which is financed by three departments – DFID, the Department of Energy and Climate Change (DECC), and Defra – supports climate investment funds, works with the private sector and finances adaptation projects in developing countries (DfID *et al.* 2011b).

8 To some degree, diplomacy has of course always included attempts to steer behavior of other countries in desired directions. But the sole orientation towards an uncertain future is a distinctive feature of a (pro-)active diplomacy approach.

9 As Erik Swyngedouw (2010) remarks, there is nothing foundational in 'nature' that could be sustained. So, sustainability is an empty signifier whose reference object that shall be sustained is defined through discursive struggles.

10 The UK Climate Act required every governmental department to conduct a sector-specific impact assessment as well as to develop an adaptation strategy for its field of responsibility (DECC 2008).

11 The CDP project is a non-governmental, non-profit organization that collects and publishes data on the climate-related environmental performance of voluntarily participating businesses and cities. The rationale behind the CDP supply chain program is to "extend awareness of an organization's [in this case the MOD's, author's note] carbon footprint [...] to include climate change risks and opportunities across the supply chain" (CDP 2015).

References

Adey, Peter and Ben Anderson. 2012. "Anticipating emergencies: technologies of preparedness and the matter of security." *Security Dialogue* 43 (2): 99–117.

Allen, John. 2011. "Powerful assemblages?" *Area* 43 (2): 154–157.

Anderson, Ben. 2010. "Preemption, precaution, preparedness: anticipatory action and future geographies." *Progress in Human Geography* 34 (6): 777–798.

Aradau, Claudia. 2014. "The promise of security: resilience, surprise and epistemic politics." *Resilience* 2 (2): 73–87.

Bettini, Giovanni. 2014. "Climate migration as an adaption strategy: de-securitizing climate-induced migration or making the unruly governable?" *Critical Studies on Security* 2 (2): 180–195.

Bigo, Didier. 2006. "Protection: security, territory and population." In *The politics of protection: sites of insecurity and political agency*, edited by Jef Huysmans, Andrew Dobson and Raia Prokhovnik, 84–100. London: Routledge.

Boas, Ingrid. 2014. "Where is the South in security discourse on climate change? An analysis of India." *Critical Studies on Security* 2 (2): 148–161.

Boas, Ingrid. 2015. *Climate migration and security. Securitisation as a strategy in climate change politics.* New York: Routledge.

Cabinet Office. 2008. *National security strategy of the United Kingdom: security in an interdependent world of work.* Crown Copyright Cm 7291, Norwich: Cabinet Office.

Cabinet Office. 2009. *The national security strategy of the United Kingdom update 2009: security for the next generation of work.* Crown Copyright Cm 7590, Norwich: Cabinet Office.

Cabinet Office. 2010a. *Securing Britain in an age of uncertainty. The strategic security and defence review of work.*" Crown Copyright Cm 7948, London: Cabinet Office.

Cabinet Office. 2010b. *A strong Britain in an age of uncertainty: the national security strategy of work.* Crown Copyright Cm 7953, London: Cabinet Office.

Cavelty, Myriam Dunn, Mareile Kaufmann and Kristian Søby Kristensen. 2015. "Resilience and (in) security: practices, subjects, temporalities." *Security Dialogue* 46 (1): 3–14.

CDP, Carbon Disclosure Project. 2015. "Supply chain." Accessed May 4, 2015. www.cdp.net/supplychain.

Chandler, David. 2014. "Beyond neoliberalism: resilience, the new art of governing complexity." *Resilience* 2 (1): 47–63.

Corry, Olaf. 2014. "From defense to resilience: environmental security beyond neoliberalism." *International Political Sociology* 8 (3): 256–274.

DCDC, Development Concepts and Doctrine Centre. 2010. *Global strategic trends – out to 2040.* London: Ministry of Defence.

DCDC, Development Concepts and Doctrine Centre. 2014. *Global strategic trends out to 2045.* London: Ministry of Defence.

DECC, Department of Energy and Climate Change. 2008. *Climate Change Act 2008.* Accessed April 30, 2015. www.legislation.gov.uk/ukpga/2008/27/contents.

Defra, Department for Environment, Food and Rural Affairs. 2012. *The UK climate change risk assessment 2012 evidence report.* London: Defra.

Defra, Department for Environment, Food and Rural Affairs. 2013. *The national adaptation programme: making the country resilient to a changing climate.* London: Defra.

Deleuze, Gilles and Felix Guattari. 1987. *A thousand plateaus: capitalism and schizophrenia.* London: Burns & Oates.

DfID, Department for International Development, Foreign and Commonwealth Office FCO, and Ministry of Defence MoD. 2011a. *Building stability overseas strategy.* London: DfID.

DfID, Department for International Development, Foreign and Commonwealth Office FCO, and Ministry of Defence MoD. 2011b. *UK international climate fund: tackling climate change, reducing poverty.* London: DfID.

Dillon, Michael and Luis Lobo-Guerrero. 2008. "Biopolitics of security in the 21st century: an introduction." *Review of International Studies* 34 (02): 265–292.

Duffield, Mark. 2011. "Total war as environmental terror: linking liberalism, resilience, and the bunker." *South Atlantic Quarterly* 110 (3): 757–769.

Evans, Brad and Julian Reid. 2013. "Dangerously exposed: the life and death of the resilient subject." *Resilience* 1 (2): 83–98.

FCO, Foreign and Commonwealth Office. 2006. *Active diplomacy for a changing world: the UK's international priorities of work.* Cm 6762, Norwich: FCO.

Foucault, Michel. 1988. *Technologies of the self: a seminar with Michel Foucault,* edited by L. H. Martin, H. Gutman and P. H. Hutton. London: Tavistock.

Foucault, Michel. 2007. *Security, territory, population: lectures at the Collège de France, 1977–1978,* edited by Michel Senellart. Basingstoke, UK: Palgrave Macmillan.

Government Office for Science. 2011a. *Foresight: migration and global environmental change. Future challenges and opportunities.* London: Government Office for Science.

Government Office for Science. 2011b. *Foresight: the international dimensions of climate change. Final project report.* London: Government Office for Science.

Hagmann, Jonas and Myriam Dunn Cavelty. 2012. "National risk registers: security scientism and the propagation of permanent insecurity." *Security Dialogue* 43 (1): 79–96.

Harris, Katie. 2012. *Climate change in UK security policy: implications for development assistance?* London: Overseas Development Institute.

Kaufmann, Mareile. 2013. "Emergent self-organisation in emergencies: resilience rationales in interconnected societies." *Resilience* 1 (1): 53–68.

Kavalski, Emilian. 2009. "Timescapes of security: clocks, clouds, and the complexity of security governance." *World Futures* 65 (7): 527–551.

Kavalski, Emilian. 2011. "From the Cold War to global warming: observing complexity in IR." *Political Studies Review* 9 (1): 1–12.

Luke, Timothy W. 1999. "Environmentality as green governmentality." In *Discourses of the Environment*, edited by Éric Darier, 121–151. Oxford: Blackwell.

Mabey, Nick. 2011. *Degrees of risk: defining a risk management framework for climate security*. London: Third Generation Environmentalism Limited.

Mayer, Maximilian. 2012. "Chaotic climate change and security." *International Political Sociology* 6 (2): 165–185.

Methmann, Chris. 2014. "We are all green now: hegemony, governmentality and fantasy in the global climate polity." Institute for Political Science, University of Hamburg.

Methmann, Chris and Angela Oels. 2015. "From 'fearing' to 'empowering' climate refugees: governing climate-induced migration in the name of resilience." *Security Dialogue* 46 (1): 51–68.

Methmann, Chris P. and Delf Rothe. 2012. "Politics for the day after tomorrow: the political effect of apocalyptic imaginaries in global climate governance." *Security Dialogue* 43 (3): 323–344.

Methmann, Chris P. and Delf Rothe. 2013. "Apocalypse now! From exceptional rhetoric to risk management in global climate politics." In *Interpretive approaches to global climate governance: (de-)constructing the greenhouse*, edited by Chris P. Methmann, Delf Rothe and Benjamin Stephan, 105–121. London: Routledge.

MoD, Ministry of Defence. 2010a. *Defence in a changing climate* London: MoD.

MoD, Ministry of Defence. 2010b. *MOD climate change delivery plan*. London: MoD.

MoD, Ministry of Defence. 2010c. *MOD climate change strategy 2010*. London: MoD.

MoD, Ministry of Defence. 2010d. "Section 7: method of climate impacts risk assessment." in *MOD sustainability and environmental appraisal tool handbook*. London: MoD.

MoD, Ministry of Defence. 2011. *Sustainable development strategy – a sub-strategy of the strategy for defence 2011–2030*. London: MoD.

Oels, Angela. 2012. "From the 'securitization' of climate change to the 'climatization' of the security field: comparing three theoretical perspectives." In: Jürgen Scheffran, Michael Brzoska, Hans G. Brauch, Michael Link and Janpeter Schilling, editors, *Climate change human security and violent conflict: challenges for social stability*, 185–206. Berlin: Springer.

Oxford dictionary online. 2015. "Diplomacy." Accessed April 30, 2015. www.oxforddictionaries.com/definition/english/diplomacy?q=diplomacy.

Paterson, Matthew and Johannes Stripple. 2013. "My space: governing individuals' carbon emissions." In *Interpretive approaches to global climate governance: (de)constructing the greenhouse*, edited by Chris Methmann, Delf Rothe and Benjamin Stephan, 139–151. London: Routledge.

Price-Smith, Andrew T. 2009. *Contagion and chaos: disease, ecology, and national security in the era of globalization.* Cambridge, MA: MIT Press.

Reid, Julian. 2012. "The disastrous and politically debased subject of resilience." *Development Dialogue* 58: 67–80.

Rothe, Delf. 2011. "Managing climate risks or risking a managerial climate? State, security and governance in the international climate regime." *International Relations* 25 (3).

Smith, Dan and Janani Vivekananda. 2007. *A climate of conflict: the links between climate change, peace and war.* London: International Alert.

Swyngedouw, Erik. 2010. "Impossible sustainability and the post-political condition." In *Making strategies in spatial planning,* edited by Maria Cerreta, Grazia Concilio and Valeria Monno, 185–204. Heidelberg: Springer.

Vaughan-Williams, Nick. 2010. "The UK border security continuum: virtual biopolitics and the simulation of the sovereign ban." *Environment and Planning-Part D* 28 (6): 1071.

Vivekananda, Janani, Janpeter Schilling and Dan Smith. 2014. "Climate resilience in fragile and conflict-affected societies: concepts and approaches." *Development in Practice* 24 (4): 487–501.

Walker, Jeremy and Melinda Cooper. 2011. "Genealogies of resilience. From systems ecology to the political economy of crisis adaptation." *Security Dialogue* 42 (2): 143–160.

8 Conclusion – or how to mitigate a climate of complexity

In the introduction to this book I outlined a twofold paradox: on the one hand, while scientific facts about the security implications of climate change have been far from reliable, the perception of climate change as a security issue seemed to have become more and more accepted; second, this discursive shift had not yet led to the adoption of exceptional measures at the political level. And as I further argued, the one theoretical approach that would be best equipped to explain the first part of the puzzle, the securitization framework of the Copenhagen School, struggles as an explanation when it comes to the second part. These observations made up the starting point for the intellectual and analytical journey undertaken in this book. Now it is time to consider the results of this journey, to reflect on experiences undergone and to come to the most important question of all: what did we learn?

A travelogue

We started the quest for a theoretical framework capable of explaining the outlined paradox in Copenhagen – with the so-called Copenhagen approach of securitization (Chapter 2). It soon became apparent that the theoretical critique of the Copenhagen School in the literature coincided with the characteristics of the climate change case. In search of a theoretical alternative, we followed the critical voices of Bigo, Balzacq *et al.* to Paris – the origin of the so-called Paris School, which developed a sociological, practice-oriented version of securitization. However, irritated by the antagonistic worldview that characterized Paris – drawing a clear line between a 'poststructuralist' and a 'sociological' approach of securitization – I argued to return to Copenhagen to revise the Copenhagen School's framework from within. Yet we did not return without taking Foucault's discourse theory as a souvenir. Through a continuous to-and-fro movement between Copenhagen and Essex – home of Laclau and Mouffe's poststructuralist discourse theory – carrying Foucault with us always, step-by-step a comprehensive and consequent poststructuralist theory of securitization could be carved out, the result of which can be seen in the synthesis part of Chapter 3.

Equipped with this revised framework I took you with me on a series of empirical short trips to find answers to the outlined puzzle. This second, analytical journey led first to New York and the UN – the main stage for a discursive struggle on climate change and security at an international level (Chapter 5). Here we did not merely remain in the main halls of the UN, the General Assembly and the Security Council, but also investigated the corridors, sideways and backrooms of the global governance system for possible traces of climate security discourse. The traces even led to the rooms of diverse climate change scientists and security experts, whose practices had a considerable impact on the establishment of a genuine climate security discourse and its concrete content and form.

In the next analytical step, I took you to the shores of the Mediterranean Sea (Chapter 6). What one could quickly recognize was that a fundamental rift marked this place: while for the EU the region clearly marked a future climate security hotspot, there was not much resonance for these storylines in the southern Mediterranean. The case presented a paradigmatic example to study the means of a discursive dissemination of climate security storylines: these were exported from different European actors to the MENA region. We came across different mechanisms of such a *securitization export*. Moreover, we could observe a quite fascinating phenomenon: different existing policy fields, such as the management of trans-Mediterranean migration, renewable energy, and the fight against desertification and water scarcity, were *increasingly translated* into the language of climate security.

Last, we arrived in the UK, a clear forerunner in climate security discourse, to study the forms of institutionalization and policy implications. It turned out that climate security rationales were so all-embracing that their manifestations could hardly be located. The first hint led to the central government, which integrated climate change into its overall approach to national security. It became apparent that one could trace the discourse back into many government departments such as the MoD, DfID and the FCO. All of these agencies had adopted climate change and security rationales into their established routines and operations. But the journey went even further. The risk-based all-hazard approach to security of the UK proved to be so all-embracing that even the individual household as well as overseas developing countries were integrated into the management of potential climate risks. Table 8.1 summarizes and compares the different cases. The table outlines the respective research focus and units of analysis in the respective cases, the specific characteristics of the respective political level compared to the others and the empirical findings in terms of different logics of securitization.

The different logics of securitization revisited

If you recall the retroductive research logic of a circular movement between theory development and empirical testing, you can imagine the

Table 8.1 Comparing the results of the case studies

Case	*Aim of the chapter*	*Case specific characteristics*	*Logics of securitization*
The international	Reveal the general *discursive structure* of climate security Study *the emergence of a climate security discourse coalition* Analyze its political repercussions in *global climate governance* (international or UN level)	Opposing storylines drawing on the same discursive vocabulary No militarization observable; instead shift towards resilience and a *post-interventionist paradigm* Post-political populism: humanity, the global or planet as referent objects Þ *High degree of discourse structuration but rather low degree of institutionalization*	*Securitization as discursive struggle* Securitization as a *reconfiguration of the governmental landscape* of climate governance = no novel policies but only a reinforcement and partial recombination of existing political technologies
The Mediterranean	Show how the climate change security storylines become disseminated Study how climate change and security impacted *development policy, including EU practices of region-building*	Focus on spatial constructions in climate change and security discourse Differences in the discursive structuration: high in case of the EU; low in case of MENA countries Rationalities and practices of climate security discourse in the field of EU's Mediterranean politics	*Securitization as an export:* practices of European actors to spread climate security storylines (and related political practices) in the southern Mediterranean *Securitization as a translation*: other issues become rearticulated in terms of climate security
The United Kingdom	Analyze in detail the *institutionalization* of the discourse where it is most probable Study the *climatization* of security and development policy	Climate security constitutes part of an *all-hazards* approach to security Climate security becomes embedded within the broader narrative of the *UK in an age of uncertainty* *Governmentality of sacrifice*: self-government integrated into security policy Þ *High degree of discourse structuration and institutionalization*	*Climate security mainstreaming:* institutional actors adopt climate security rationales within their established routines *Institutionalization as assembling*: climate security assemblage of different practices and rationales that do not follow a broader or overall logic

effort of the outlined journey not in terms of kilometers traveled but in terms of ink and brainpower. Yet what has been the merit apart from nice memories, which are of less interest to others? Let me come back to my initial puzzle. First, why and how could a security framing of climate change become so prominent despite all the doubts about such linkages? There is not one single answer but several to this question. The answers include both theoretical considerations and the empirical investigations made in this book.

First, a security framing of climate change became prominent as a result of a longer discursive struggle. This discursive struggle grew out of a long-standing discourse on environmental security and led to the establishment of a transnational discourse coalition revolving around different storylines of climate security. Starting with the articulations of a few academics and ministerial officials in the early 2000s, the discourse coalition won over more and more actors from different subject positions in the course of the mid-2000s: climate celebrities like Al Gore or Ban Ki-moon, victims of climate change such as the SIDS, industrialized countries such as the UK as forerunners of climate security, humanitarian and environmental NGOs as well as security related think tanks. While many of these voices remained skeptical or at least cautious about climate security linkages, a range of consensual nodal points helped pacify this field of contestation. Such pacifying discourse elements were for example the concept of climate change as a threat multiplier, the nodal point of climate refugees linking humanitarian, security and development concerns, and particularly emblematic cases of climate security such as Darfur, Bangladesh and the fate of the SIDS. In sum, two developments made up the discursive hegemony of a climate security framing: first, the establishment of a discourse coalition around climate security storylines ranging from national leaders to security professionals and humanitarian and environmental NGOs; and, second, the pacification of the discursive field through consensus-producing concepts and storylines.

A second answer to the question would argue that it has been the practices of multiple actors working in the field(s) of climate change and security that made the security framing prominent. These often unnoticed, rather mundane practices of climate scientists, risk analysts and security professionals are for example computer modeling, satellite monitoring and remote sensing, scenario development and planning, war games and role plays, vulnerability mapping, emergency exercises and assembling of risk registers. Third, climate change and security rationales were able to become so prominent because they were actively exported by Northern actors in order to induce an awareness of the climate change threat in the South. This could be clearly shown in the case of the Mediterranean. For example, I outlined how initiatives such as the Mediterranean Security Initiative used expert workshops as well as methods such as participatory mapping exercises to raise awareness of those vulnerable to climate change in the southern Mediterranean. At the same time, joint operations of experts, for example on civil protection, between European and MENA actors allowed the production of a joint threat perception. And lastly, I showed in the empirical study that climate change and security discourse was able to become so strong because it managed to integrate and translate a whole range of other political issues and problems. This holds true for other environmental problems, such as desertification, water security or the question of conflicts through resource scarcity. But this also holds true for the discourse on irregular migration, political radicalization and other non-traditional security threats.

These four explanations are *not mutually exclusive* and none of them is a totalizing logic, i.e. none of them alone could explain the outlined puzzle. I want to understand them as *different logics of securitization*, which, depending on the respective case, can be more or less present.

Let me now turn to the second part of the puzzle: why did this securitization of climate change not translate into extreme or exceptional measures in the sense of the Copenhagen School? The answer is quite straightforward: because the political implications depend on the respective case. Which factors influence the political implications of a securitization process in a particular case? On the one hand, the concrete discursive structure – or, more precisely, the particular *articulation of the antagonism* – has implications for the policy instruments adopted. In the case of climate change, it was constructed as an overall enemy of humanity. This macro-securitization at the global level constructs the whole of humanity as a homogeneous social space, which is then governed through a logic of risk and resilience based on the established mechanisms of the UNFCCC. This threat is simply so overarching and at the same time so spectral that it cannot be fought by exceptional measures of any political leader but can only be fought through the everyday practices of everybody.

On the other hand, the concrete institutional context influences the political impacts of the discourse. As I showed in the case of the UK, the adoption of climate change and security rationales by governmental actors mainly followed a logic of climate security mainstreaming. Officials of the FCO or DfID drew on climate change as an already securitized issue and then integrated it into their own established routines and operations. The result was what I called an institutionalization through assembling.

There are two fundamentally different ways that the storylines articulated by a discourse coalition could materialize into policies or institutions. The first would be based on the establishment of a consensual problem definition and, deduced from that, a common political will of the discourse coalition. This common will could then be turned into a coherent and comprehensive policy program on climate security in which every part of the actor coalition would play a particular role, subjected to an overall governmental logic. Exactly the opposite is the case in climate security mainstreaming. Here, each actor in the discourse coalition translates the shared storylines into policies or institutions on the basis of her own institutional routines. The result, then, just as in the case of the UK, is a complex assemblage of practices and policies, which are loosely connected through their common orientation towards climate security but do not follow an overall total logic (and hence might even be counterproductive or contradictory).

Exceptional security or mundane resilience? Yes, please!

This book addresses not only the debate on the securitization of climate change and its implications but also the theoretical debate on securitization, risk and resilience. I would argue that my main contribution here has been to tear down conceptual walls. This began with my decision to resist the attempt to formulate an alternative approach in opposition to the Copenhagen School's framework. Instead of being a bridge between the different schools of securitization,[1] my book can be understood as a wrecking ball to deconstruct the outer defence walls of these theoretical fortresses. It is my deepest conviction that one should not oppose the language/discourse-oriented approach to securitization to a sociological variant, as for example the Paris School did (Balzacq 2010: 1–3). The importance of language as speech and writing on the one hand and the impact of multiple practices and even material objects on the other for the social constitution of reality are two sides of the same coin and must not be taken as opposing approaches. In the same way, the book sought to tear down the massive defense walls that have been built between risk and security-oriented approaches (see also Corry 2012). Both securitization literature and Foucault-inspired research on risk in security studies take these logics as two different animals – for example as geopolitical versus biopolitical dispositif of security (Dillon 2007: 5). Yet, as I tried to show in all three empirical chapters, these different political rationalities and practices are often intrinsically linked. Hence, I would argue that we should not make a conceptual a priori distinction between different *dispositifs* of security but assume the existence of a security continuum from the global apocalypse to insecurities of the individual (Bigo 2000: 323–324).

In addition, I want to argue *against unidirectionality* in the study of securitization processes. In the original framework of the Copenhagen School, securitization is a unidirectional force. Depending on where one looks, through the prism of the Copenhagen School's framework one sees either that an issue is taken out of its 'proper' context and brought into the field of security or an expansion of the security field that colonizes ever more political issues. Angela Oels has argued that in the case of climate change one should take this perspective upside down, as what we can see here is rather a 'climatization' of the field of development, security and migration management (Oels and Brzoska 2012). Although this certainly is an important argument I would claim that we should even go one step further. Oels's perspective still adheres to the unidirectionality inherent to the Copenhagen School approach: now it is climate discourse that expands and thereby colonizes ever more policy fields. Yet in this book I showed that this represents only one of many possible discursive interactions. Securitization should rather be understood as interplay between and mutual imbrications of specialized discourses that can follow different directions. In this respect there was both a transfer of security rationales to the climate change discourse and at the same time a spillover of concepts from climate change discourse into the security field. The concept of resilience is a good example: while it developed as part of environmental thinking in ecosystems research (Holling 1973) it was soon taken up in US national security strategy (Walker and Cooper 2011: 152–154) only to be reintroduced into environmental discourse through a growing focus on climate impacts, vulnerability and adaptation (Bahadur *et al.* 2010). In the complex interplay between different discourses such as those on climate change, development or security, often one cannot tell the exact source of rationalities or the direction of their travel.

This assumption leads me directly to another problem: scholars have often identified different shifts in security thinking and policy, from defense to resilience, from security to risk, from risk to anticipation, from prevention to preparedness, etc. I take this to be partly driven by the disciplinary need to *identify the next big trend* and the requirement to sell and advertise research findings in a competitive academic economy. Yet I want to resist this temptation to sell my result in terms of an overall trend. The topological analytical perspective that I adopted from Collier (Collier 2009) in this book resists any arguments about totalizing shifts. Instead of a shift towards something new, the political implication of a securitization of climate change represents a *complex and creative reconfiguration of the existing*. And it stresses the multifaceted nature of climate change and security discourse, which means that any sort of overall trend or logic in climate security thinking is improbable.

Contribution to the literature on climate change and security discourse

The combination of a detailed synchronic discourse analysis with a broader genealogical perspective in the present book adds to the recent debate on the securitization of climate change in important ways and helps in painting a much more comprehensive picture of climate change and security discourse. I was able to show that the securitization of climate change cannot be separated from but rather directly links up with other discursive changes in the past: the broadening of security and the environmental security debates in the 1990s, the securitization of desertification, and the development–security discourse that became prominent after 9/11. The inclusion of contextual discourses, i.e. related discourses such as development and security to which climate security storylines link up, represents a further merit of this book. Thereby, I was able to show that securitization does *not simply* mean *taking an issue out from political discourse* into the realm of security discourses as suggested by the Copenhagen School but must be understood as a *creative interplay* between different specialized discourses.

As the case study of the UK particularly exemplifies, this can lead to productive spillovers of certain policy models (such as risk and resilience) from one policy field to the other (e.g. from disaster planning to peace-building to national security and climate adaptation practices) and hence to a convergence of policies between certain fields (such as development and climate change policy). Finally, only by understanding securitization as a process of *discursive struggle*, as this book does, does one become able to assess the present strength and political attraction of climate security. Through this perspective I was able to show how different storylines and actor coalitions were competing in the discursive field and also through which discursive strategies the dominant discourse was able to absorb and incorporate critical voices. Such a diachronic perspective revealed how existing policies such as the concept of Euro-Mediterranean disaster planning activities were rearticulated in terms of the climate security discourse (Chapter 6). And such an approach enabled one to notice how climate security was embedded within broader security rationalities held together by a complexity narrative (Chapter 7).

Studies on the securitization of climate change informed by the Copenhagen School in the past were to some extent haunted by a *quest for exceptional measures*. All contributions in this literature basically agree that we do not see any exceptional measures in the sense of the Copenhagen School. Contributions discussed, for example, whether existing policies could be regarded as being exceptional or searched for reasons of the non-existence of exceptional measures intrinsic to the climate change case – this included some of my own contributions in the past (Rothe 2012). In this respect studies have blamed the non-existence of exceptional measures on the variety of competing security discourses (see e.g. Brzoska 2009), have argued that securitization articulations are only strategic in nature and real politics are influenced by other factors (Brzoska 2009: 144), and that political effects will possibly materialize in the future – if for example climate conflict storylines are used to make military interventions in crisis areas more palatable (Hartmann 2013). Hence, based on the findings of this book I would argue that one can better explain the political impacts of a securitization of climate change if one abandons the category of exceptional measures and understands securitization as a process of reconfiguration.

Both an approach that understands securitization as a causal mechanism (Guzzini 2011) as well as a constitutive approach stressing the constitutive role of performative securitizing acts obfuscates certain political implications of climate security that are already there. This is because in both perspectives that which causes or constitutes something must exist prior to its effect. In both cases, a temporal structure is constructed that blurs the often messy and multiple nature of securitization processes (Huysmans 2014: 109). This book hence follows a different interpretation – resisting the either/or perspective by taking up a processual approach to securitization. Policy implications of securitization might appear in this process of securitization itself – for example the growing dominance of practices of risk assessment, resilience, and other forms of technocratic management.

That being said, I am certainly not claiming to have reinvented the wheel here. Rather, I would explicitly locate myself within the literature drawing on risk and a Foucault-inspired understanding of security. All existing studies, however, share their focus on macro-discourses or governmentalities, studying the rationalities, practices and subjectivities that are visible in major documents and articulation of the discourse. The present book adds important insights to this literature by showing, on the one hand, how rationalities and practices of a climate risk management are backed up and promoted by a certain discursive deep structure (see Chapter 4) as well as storylines and narratives that permeate climate security discourse (see Chapter 5). On the other hand, this analysis provided an insight into the functioning of climate risk management at a micro-level, with an attention to detail that was hitherto missing in the debate. By incorporating three political levels from the international to the regional and national, the book showed the variety of forms that "promoting climate resilience" can take. Climate resilience management is an epistemological model that informs – and serves as an envelope for – a whole variety of political practices and policies: from international mitigation and adaptation practices within the institutions of the UNFCCC and the Kyoto Protocol, to the promotion of practices of self-management and sacrifice or the concept of an active climate diplomacy as in the case of the UK.

Criticizing oneself – criticizing others

With my book I also tried to relate to the question about what it means to be critical. Broadly speaking, I am critical of the form in which criticism is often expressed in scientific and also everyday discourse. This is because criticism here always points to the failure or wrongdoings of others. Criticism then becomes a self-defending, knee-jerk reaction, in which the contradictions of a system and the miserable situation of a subject is ascribed to a *personified other*. In this form of criticism, which is particularly strong in parts of the European left (in the right anyway), a few bankers or managers are responsible for the failure of capitalism, the US is responsible for all the violence in the world and a few oil companies responsible for the ecological crisis. I would argue that a Copenhagen School approach ultimately supports such a form of externalized criticism, if it focuses on the practices of political elites that securitize particular issues in order to break with established rules. Certainly, people have different capabilities to shape the social and political world but a social system is always reproduced by all of its members. This holds especially true for the social system that produces a global climate crisis. We are all contributing to this crisis. The paradox is that it is perfectly possible to work and to write on the gravest implications of this problem and at the same time stick to all the routines and habits that we know are producing this catastrophe. I want to stress a form of criticism, which always includes self-criticism. In this respect I have to confess that, although I try not to do so, I contribute to many of the problems and looming threats described in this book.

Stressing the importance of self-criticism also has implications for a normative approach to securitization. This is because, as a researcher, I am never able to take up a position outside of the discourses that I describe. Simply by participating in a discursive struggle, by writing a book about the growing securitization of climate change, there is always the risk of producing undesired effects. For example, my arguments that are skeptical of a security take on climate change or that conceive the world as discursively constructed might be taken up by climate change skeptics to argue against the need to adopt progressive climate policies. This does not mean that we should not engage in such discourses anymore – quite the contrary – but I want to stress that one should be aware of the responsibility and the risks that this endeavor involves.

Critical questions about this book

A first critical question one could ask about this book refers to the theoretical framework that I developed. One could ask if it was really reasonable to turn the decidedly slim securitization framework into such a complex theory. I certainly took this question seriously. I am well aware of the fact that Buzan *et al.* decidedly kept their framework slim and reduced it to a few factors such as the securitizing actor, the move and the audience. Yet I am absolutely convinced that it was not only reasonable but also necessary to place securitization on a more comprehensive and deep theoretical foundation. As attractive as it might seem, such an application-oriented framework in a 'handbag format' will always fall short of social reality. It is not only the numerous critical voices in the literature but also the problems of the framework in the case of climate change that proved this. To understand and explain the complexity of climate security discourses one needs more than a slim perspective: we have to study this subject from various angles and have to dig deeper in order to be able to grasp its complexity.

A second critical question refers to the relation between the theoretical framework and the empirical findings. When discussing my research project, several colleagues asked why I would actually need the securitization perspective. Would it not be possible to conduct the same empirical investigation simply by drawing on, for example, Foucault's discourse theory? There are basically two answers to this question: one concerns the securitization approach and the other my empirical analysis. On the one hand, I would argue that, despite all the justified criticism, securitization still represented a brilliant idea. This applies to a discursive reading of security and also to the central role of political antagonisms in the legitimization of certain policies. Additionally, the recently outlined concept of macro-securitizations proved to be a very useful concept in the study of the climate change and security discourse. Furthermore, the Copenhagen School framework exhibited so many linkages to poststructuralist theory that it represented a perfect starting point for the development of a post-structuralist theory of security. With regards to the empirical cases studied in this book, let me say that although it departs from a realist or conventional understanding, it is all about security! It is not simply about the construction of risk or danger but also about the very way we think about and act upon *security*. Hence, which theory, if not a securitization approach asking for the expansion and broadening of security discourse, should be applied to such cases?

A third and last question refers to the status of the knowledge acquired in this book. A very critical, and certainly every neo-positivist, reader would probably ask how I could seriously claim to have explained the growing influence of security thinking on climate change. Was this book not merely a very detailed and thick description of a discursive development and forms of its institutionalization? My answer would be: Yes, certainly! And no, of course not! As I outlined in Chapter 4 of this book I promote an approach to science that not only requires description and explanation but also questions the boundary between these two logics of research. To explain, according to this perspective, always requires understanding – to make sense of something. I hope that I was able to make clear that there are *no primary causes* of a securitization of climate change – not an initial actor who decided to construct climate change as a security issue. Rather, I presented a complex assemblage of different developments and factors. And taken together the complex interplay between these factors represents *the cause* for the fact that climate change became constructed as a security threat. In this respect, yes, I have explained the puzzle outlined in the introduction, even if this might not be a satisfying explanation in the view of neo-positivist scholars.

Securitization, normativity and critique

So what does it mean to be critical in the study of securitization processes? Criticism here refers to the question of whether scholars should actively engage in a desecuritization of climate change, e.g. by opposing a security framing of climate change or by promoting alternative problem definitions. The Copenhagen School approach included a clear affirmation of desecuritization as a normative goal of critical scholarship (Aradau 2004: 389; Buzan *et al.* 1998: 29). Against the backdrop of the analysis presented in this book, such an endeavor appears at the very least doubtful, if not problematic. The securitization of climate change, it was shown, cannot be regarded as a linear process that could somehow be reversed or rolled back. The notion of desecuritization once more shows the idea of unidirectionality underlying the approach of the Copenhagen School. The possibility of simply desecuritizing is informed by the notion that there would be some primordial and proper meaning of any political problem that could somehow be restored. However, as the case of climate change discourse shows, securitization is only one episode in an infinite and ongoing negotiation of meaning. In other words there is no 'climate change proper.' Desecuritization, following the perspective developed here, could only mean participating in the discursive struggle to establish alternative framings and problematizations. However, arguing that there is no genuine or right problematization of an issue also implies that alternative framings to a climate security discourse – e.g. an economic discourse – are not necessarily normatively superior (McDonald 2015).

A further insight produced by this analysis is the impossibility of taking up a position outside of the discourse. The claim for desecuritization implies that the critical scholar herself would not participate in the securitization process and could quasi form the opposition. However, the analysis in Chapter 5 showed that critical academic voices from the very beginning were included in the climate change and security discourse. It is precisely the ability to include and rearticulate critical voices that makes up the hegemony of any discourse. Even the securitization approach of the Copenhagen School is well known by many actors participating in the climate change and security discourse. For example, my interviews showed that nearly all interviewees were aware of the theoretical framework of the Copenhagen School or at least of the critique that a securitization of climate change could have negative effects. These voices explicitly refer to the dangers of a militarization of climate politics – without questioning the relation between climate change and security in general. This shows that it is not possible to strategically influence the discourse as an outsider per se, as academics are always already part of the securitization discourse itself.

Given these problems with a desecuritization strategy I want to argue for a normative approach that focuses on critique as a form of 'showing up' or 'making visible.' This is a position of *immanent critique* (Antonio 1981; Nunes 2012: 352). Rather than criticizing climate security discourse from an outside and already given standpoint, the researcher should critically engage with the respective discourse to develop a critique from within. It is a critical position that tries to make visible the contingency of the dominant security framing by revealing its origins and genealogy, and its contradictions and possible negative implications. In this respect there is nothing objectively or universally negative about the securitization of an issue. Instead the securitization of climate change must be judged on the basis of what it actually does. And by taking a deeper look at the dominant climate change discourse we can identify several points that can be seen critically: the construction of the vulnerable as helpless victims; the technocratic nature of risk governance, in which risks and measures to cope with them are defined by a few experts; the mechanistic notion of causality, according to which political violence occurs as a product of a range of push and pull factors; the construction of climate change as an external enemy, which blurs existing inequalities and antagonisms in international politics; and the fact that the discourse is almost exclusively made up of voices from the global North. So, if there is one essential feature of the securitization of climate change that could be identified and criticized, it is the conservative nature of climate security discourses and practices. For the primary concern of all articulations of climate security discussed in this book is to preserve a certain state of the 'environment' and in particular to sustain the existence of our capitalist economy. The case of climate change adds further evidence to the argument that " 'Security' has always been about mitigating the contradictions of capitalism and profiling the conditions necessary for its reproduction" (Dalby 2015: 6).

From my perspective there are hence two minimal requirements for a critical perspective on the securitization of climate change. First, a critical position is a position which highlights the contingency of a security discourse and which points to the existence of political alternatives. Instead of desecuritization, it argues for politicization understood as the promotion of open debate and contestation (Aradau 2004). But what could such alternatives then look like if there is not a proper desecuritized discourse on climate change? For example, one could try to change the climate change and security discourse itself. By letting those that are most affected by the negative side effects of climate change speak, one could try to change the Northern bias of contemporary climate security discourse. For example, rather than conceptually discussing the question of whether climate change represents a source of migration or not, one could simply listen to those people that are already on the move. Another possibility would be to engage in discursive struggles about key concepts such as human security. It is not as if there would be no progressive political potential in such concepts, for example if human security were really to follow emancipatory goals in line with critical theory (Booth 1994). Resilience could also be understood as a concept that focuses on the needs of the people and increasing their wellbeing (Methmann 2014: 250). The problem then is simply that such concepts are defined and interpreted against the backdrop of hegemonic discourses (Booth 2007: 326). But if more and more people engage in critically discussing and rearticulating such concepts, it might be possible to change their meaning so that they serve the wellbeing of people.

A second qualification to a critical approach to climate security would be one that challenges the conservative nature of the discourse by stressing the possibility of political progress and transformation. This would be an approach that stops romanticizing the present and invoking fears about the future but one that defines possible visions of desirable forms of political and social organization in the future as well as ways to realize them. As Julian Reid puts it for the case of climate-induced migration:

> Once we step outside of the biopolitical imaginary that accounts for its pathologization, on account of the ways in which it sees the phenomenon of climate-induced migration as a potential threat to the conservation of global liberal order, so we are freed to find other ways of imagining and thus valuing it.
>
> (Reid 2014: 207)

While this does not imply welcoming the negative effects of climate change for large parts of world's population, especially in the global south, it means taking the existence of the anthropocene as a starting point to attack the current system of sovereign nation states and to fight for a world without borders.

Note

1 See the CASE Manifesto for such an attempt (CASE Collective 2006).

References

Antonio, Robert J. 1981. "Immanent critique as the core of critical theory: its origins and developments in Hegel, Marx and contemporary thought." *British Journal of Sociology* 32 (3): 330–345.

Aradau, Claudia. 2004. "Security and the democratic scene: desecuritization and emancipation." *Journal of International Relations and Development* 7: 388–413.

Bahadur, Aditya V, Maggie Ibrahim and Thomas Tanner. 2010. *The resilience renaissance? Unpacking of resilience for tackling climate change and disasters.* Brighton: Institute of Development Studies.

Balzacq, Thierry. 2010. "A theory of securitization: origins, core assumptions, and variants." In *Securitization theory: how security problems emerge and dissolve*, edited by Thierry Balzacq, 1–30. London: Routledge.

Bigo, Didier. 2000. "When two become one: internal and external securitisations in Europe." In *International relations theory and the politics of European integration: power, security, and community*, edited by Morten Kelstrup and Michael C. Williams, 154–183. London: Routledge.

Booth, Ken. 1994. "Security and self reflections of a fallen realist." *YCISS Occasional Paper* 26, Aberystwyth.

Booth, Ken. 2007. *Theory of world security, Cambridge studies in international relations, 105.* Cambridge: Cambridge University Press.

Brzoska, Michael. 2009. "The securitization of climate change and the power of conceptions of security." *Security and Peace, Special Issue on Climate Change and Security* 27 (3): 137–145.

Buzan, Barry, Ole Wæver and Jap De Wilde. 1998. *Security: a new framework for analysis.* Boulder, CO: Lynne Rienner Publishers.

CASE Collective. 2006. "Critical approaches to security in Europe: a networked manifesto." *Security Dialogue* 37 (4): 443–487.

Collier, Stephen J. 2009. "Topologies of power. Foucault's analysis of political government beyond 'governmentality'." *Theory, Culture and Society* 26 (6): 78–108.

Corry, Olaf. 2012. "Securitisation and 'riskification': second-order security and the politics of climate change." *Millennium-Journal of International Studies* 40 (2): 235–258.

Dalby, Simon. 2015. "Climate geopolitics: securing the global economy." *International Politics,* published Online March 20, 2015.

Dillon, Michael. 2007. "Governing terror: the state of emergency of biopolitical emergence." *International Political Sociology* 1 (1): 7–28.

Guzzini, Stephano. 2011. "Securitization as a causal mechanism." *Security Dialogue* 42 (4–5): 329–341.

Hartmann, Betsy. 2013. "Climate chains: neo-Malthusianism, militarism and migration." In *Interpretive approaches to global climate governance: (de)constructing the greenhouse,* edited by Chris P. Methmann, Delf Rothe and Benjamin Stephan, 91–104. London: Routledge.

Holling, Crawford S. 1973. "Resilience and stability of ecological systems." *Annual review of ecology and systematics*: 1–23.

Huysmans, Jef. 2014. *Security unbound: enacting democratic limits.* London: Routledge.

McDonald, Matt. 2015. "Climate security and economic security: the limits to climate change action in Australia." *International Politics,* published online March 20, 2015.

Methmann, Chris. 2014. "We are all green now: hegemony, governmentality and fantasy in the global climate polity." Institute for Political Science, University of Hamburg.

Nunes, João. 2012. "Reclaiming the political: emancipation and critique in security studies." *Security Dialogue* 43 (4): 345–361.

Oels, Angela and Michael Brzoska. 2012. "From 'securitization' of climate change to 'climatization' of the security field: comparing three theoretical perspectives." In *Climate change, human security and violent conflict,* edited by Jürgen Scheffran, Michael Brzoska, Hans Günter Brauch, Peter Michael Link and Janpeter Schilling, 185–206. Heidelberg: Springer.

Reid, Julian. 2014. "Climate, migration, and sex: the biopolitics of climate-induced migration." *Critical Studies on Security* 2 (2): 196–209. doi: 10.1080/21624887.2014.943578.

Rothe, Delf. 2012. "Security as a weapon: how cataclysm discourses frame international climate negotiations." In *Climate change, human security and violent conflict,* edited by Jürgen Scheffran, Michael Brzoska, Hans Günter Brauch, Peter M. Link and Janpeter Schilling, 243–258. Berlin: Springer.

Walker, Jeremy and Melinda Cooper. 2011. "Genealogies of resilience. From systems ecology to the political economy of crisis adaptation." *Security Dialogue* 42 (2): 143–160.

Index

For Product Safety Concerns and Information please contact our EU representative GPSR@taylorandfrancis.com
Taylor & Francis Verlag GmbH, Kaufingerstraße 24, 80331 München, Germany

www.ingramcontent.com/pod-product-compliance
Ingram Content Group UK Ltd.
Pitfield, Milton Keynes, MK11 3LW, UK
UKHW022132190926
475702UK00013B/1772
9781138096530